Near-Death Experiences

4|15

Birk Engmann

Near-Death Experiences

Heavenly Insight or Human Illusion?

 Springer

Birk Engmann
Leipzig
Germany

Expanded and revised translation of the German language edition *Mythos Nahtoderfahrung*
by Birk Engmann
Copyright © S. Hirzel Verlag GmbH Co, Stuttgart, Germany (2011)

ISBN 978-3-319-03727-1   ISBN 978-3-319-03728-8 (eBook)
DOI 10.1007/978-3-319-03728-8
Springer Cham Heidelberg New York Dordrecht London

Library of Congress Control Number: 2013957431

Printed on acid-free paper

Springer is part of Springer Science+Business Media (www.springer.com)

Contents

1

Introduction

No aspect of science and research has been discussed more emotionally than the topic of near-death experiences. The media have shown continued interest in this question for several decades now. One reason may be that such experiences fall in the overlap between neurological sciences, psychology, psychiatry, philosophy, and the science of religion, something that may also explain the broad interest of the general public. The very term of *near-death* experience reveals its emotional content. It is a term which suggests knowledge of one's own existence at the dividing line between life and death. But how might we apprehend such knowledge? Could such experience be verified by scientific methods, and if so, how could it be explained?

The multitude of views and models purporting explanation already indicates that near-death research is something of a tightrope walk between rationally explainable theories and the sphere of belief. But here the traditional quarrels between science and religion are not really the focus of public attention. Quite the contrary in fact: esoteric constructions which include occultism, superstition, and all kinds of spirituality have become more and more prominent. In this context, there is an urgent need for a critical review, and indeed, one which examines the way the natural sciences can throw light on this matter. This is the main aim of the book. Chapter 2 deals with historical aspects, while Chap. 3 discusses problems with the term "near-death experiences". The rest of the book considers models for medical explanation. Technical terms are explained in an appendix at the end of the book. It is my wish that the book will stimulate new and critical views on the topic of "near-death experience". To my mind, continually bringing one's own convictions into question is a precondition, not just for scientific work, but for any intellectual progress. Moreover, in these days when confident, selfish, and showy behaviour seem to be prized, a little humility would be in order, given the imperfections of human understanding against the backdrop of our endless universe.

Despite the long history of debate and research regarding near-death experiences, several myths have nevertheless set in:

* Near-death experiences already appear in reports from ancient times, whence they bear the stamp of immemorial knowledge.
* Near-death experiences have unique characteristics which set them apart from similar experiences in other circumstances than clinical death.
* Near-death experiences are generated when brain function is completely absent.
* Near-death experiences provide a proof of the existence of god or the supernatural.
* Near-death experiences provide a proof that god or the supernatural does not exist.

These myths will be investigated in the following pages.

2

Is There Historical Evidence of Near-Death Experiences or do We Merely See What We Want to See?

2.1 The Wondrous World of the Dutch Painter

The beginning of the 16th century was a time of many problems in Europe: pestilence, the early stages of a period of climate change which we call today the Little Ice Age, and hostile armies threatening Europe from all sides. It is no surprise therefore that many people already feared the approaching apocalypse, and it is perhaps no coincidence that this period has produced some of the strangest and most mysterious paintings. One of these, which is attributed to the painter Hieronymus Bosch (born around 1450 in 's-Hertogenbosch and died in 1516, ibid.) by many but not all art historians, is still a subject of fascination today. Not only does it depict fabulous creatures, but it also deals with one of the most important issues for human beings: the question of life after death.

The central topic of the painting *Ascent of the Blessed* is the ascent of the soul into heaven, a topic that is deeply embedded in our civilization. The painting, oil on wood, was done in the years 1500 to 1504, and can be seen in the Doge's Palace in Venice. It illustrates how decedents move though a kind of tunnel, at the end of which is a very bright light. So is this artistic freedom or insight into some real experience? (Fig. 2.1)

Amazingly, this painting is the one most often reproduced in publications dealing with the topic of near-death experience. Indeed, it has already gained an iconographic significance: anyone who sees the painting already knows what the publication is about. But one should really highlight other features. To begin with, the painting is one of four panels entitled *Visions of Hereafter*. The polyptych begins with *Terrestrial Paradise*. This is followed by *Ascent of the Blessed, Fall of the Damned,* and *Hell*. It is still a matter of debate whether Bosch's panels were in any way inspired by similar paintings from the Dutch painter Dieric Bouts (born around 1410/20 in Haarlem, died in 1475 in Leuven) (Bosing 2012). In the transition from the Middle Ages to the Renaissance, the biblical account of the last judgement, which falls upon everyone

Fig. 2.1   Hieronymus Bosch: "Ascent of the Blessed", oil on wood, detail

after death and which provides all those who suffered from injustice on Earth with transcendental redress, was an important topic of artistic depiction. In this content, it is worth remembering an approach to that same theme in literature, long before Bosch's time, by Dante Alighieri (born 1265 in Florence, died 1321 in Ravenna). Making full use of artistic freedom, he travelled through hell, purgatory, and paradise accompanied by the Roman poet Virgil.

Unfortunately, too little is known about the life of Hieronymus Bosch to provide a satisfactory explanation of the *Ascent of the Blessed* on the basis of his biographical background. As Bosing (2012, p. 38) has argued, the funnel-shaped light could just have been inspired by the astrological signs of his time. Indeed, anyone who looks carefully at the painting will easily recognize six regular ring-shaped sections in the funnel. Another author, Schürmeyer (1923), emphasizes the symbolic content in all Bosch's work (Fig. 2.2):

Bosch is anything but the inventor of all the thousand variants of demons and nuisances; he certainly did not shape the well known proverbs which he had

Fig. 2.2 Self-portrait of Hieronymus Bosch or posthumous portrait attributed to Jacques Le Boucq. (Born 1520, Valenciennes; died 1573). Arras/France, municipal library

painted allegorically. He only put new ideas in painting, and the way he created them characterizes him as an artist and man. (p. 31)

2.2 If I Have not Charity, I am Nothing—Essential Questions of Our Existence

Half a millennium after Bosch, the American psychiatrist Raymond A. Moody (born 1944 in Porterdale, USA) published a collection of near-death experiences. Some of the stories seem to suggest an association with Bosch's painting. People who were clinically dead but subsequently brought back to life by reanimation also remembered a tunnel with bright light at its end. Even today many of the people concerned agree upon the artistic intentions of Hieronymus Bosch and interpret their own experiences as epiphany, that is, the appearance of a deity, or a glimpse of the afterworld. But is there another possible explanation? Could there also be a rational view? Are there scientific clues as to the nature of such experiences; experiences which may have been familiar to many generations of people?

Raymond A. Moody identified similarities in the near-death experiences of many people, with regard to both content and timing—a view which can no longer be upheld now that we possess the findings of much more research. This will be discussed further below.

According to Moody (1975) then, one can identify the following sequence of experiences in clinical death, while the exact time at which the experiences originated remains unknown. Each subject reports things they experienced, at times of varying length after the event. First, they experience unpleasant noises or a pervasive tolling or buzzing. Then the tunnel phenomenon occurs. They feel as though they are moving through a long and dark tunnel. Then they seem to leave their own body and view it from a distance. Light or an illuminated being appears, and the person "sees" spiritual versions of relatives or friends who have already died. This is accompanied by feelings of love and warmth. After that, their whole life passes by like a film.

It is surprising that, not only have the tunnel phenomenon and perception of light already become an important part of the way we imagine life near death, but even the film-like view of one's own life is considered to be a typical experience for such a state. For instance, the latter is often used in a metaphoric way by film directors when a screen hero dies. But contrary to the assumption of a strict order and clear pattern, reports on near-death experiences actually differ enormously. The following examples taken from the author's own investigations in Uzbekistan (Engmann and Turaeva 2013) will illustrate this.

A 39 year old man we interviewed had been in a car accident 3 months previously. He had been crushed inside the badly damaged vehicle and lost consciousness. When he came round, he saw doctors around him. He remembers that he wanted to go to the toilet and talked to the physicians. Afterwards he underwent an operation on the abdomen. Two days after the operation, he fell into a coma for 3 days, and was reanimated with electro-shock therapy. When he regained consciousness, he remembers hearing voices like "*on a bazaar when everything swooshs.*" He also saw his grandfather wearing a "chalat"—the traditional clothes. His grandfather said something to the patient which he did not understand. But he reports that "*others asked me why I was lying down.*" However, the patient could not remember either when the vision occurred or who these "others" were. Finally, the patient said: "*I know about such things that people see, that the soul comes out when the heart stops, but I didn't see such stuff.*" (p. 3)

Another man had a stomach operation, but apparently no reanimation. He claimed to have had a near-death experience. He saw the white coats of doctors standing in a row, but there were no people in them. There were nine of these, and the fourth one along was dancing. He drew his vision like this: □□□□□□□□□ (p. 5)

A woman who was reanimated reported that she was flying over the Earth. Everything was green and there were beautiful flowers; she had never seen such beauty in her whole life. She came to a beautiful house and met her mother, who was already dead. Her mother told her: "*You have to go back, you have children!*" (p. 5)

Having been in a state close to death and knowing that one has escaped from death at the very last moment is a shocking realization that leaves a lasting impression on anyone who has experienced such a situation. In some cases, it can seriously alter someone's personal conduct and have a significant influence on their philosophy of life. And if the latter turns around religious convictions, it often happens that near-death experiences are seen as supporting such views. In contrast to personal beliefs of this kind, many protagonists of a metaphysical explanation for near-death experiences take this as proof of the existence of the supernatural. At this point, science, personal interpretation, and religion intermingle. But there are many problems with the current erosion of the interface between science and religion when it comes to near-death experiences. Furthermore, philosophical discussions over the ages have already pointed out severe problems with the notion of a scientific proof of the supernatural. The existence of supernatural phenomena or beings can neither be proved nor refuted by scientific methods. In the case of near-death experiences, it is easy to move directly into the sphere of opinion, personal interpretation, and belief. And this sphere is not only typical for near-death experiences, but affects virtually all fields of everyday life. It includes all the questions formulated by the philosopher Immanuel Kant (born 1724, Königsberg; died 1804, Königsberg) (Fig. 2.3):

What can I know?
What ought I to do?
What may I hope?
What is man?

Thus, discussions about near-death experiences touch upon essential questions of our existence, the meaning of life, and human destiny. But near-death experiences are not the only situation that might make us think about such problems, and require a philosophical or religious approach. Other situations in life can bring about similar needs, such as the advent of a life-threatening cancer. Any such threat to one's own existence in the imminent future can

Fig. 2.3 Immanuel Kant, vintage postcard after a painting of Fritz Rumpf. (Born 1856, Frankfurt on the Main; died 1927, Potsdam)

lead one to rethink one's life. Those who have lost relatives or close friends often find themselves in a similar situation. In contrast to this, in normal everyday life, the very limitations of that life are usually pushed to the back of our minds.

And yet we all know that the most disastrous experiences of our lives are not the only cause for reflection about the basic problems of philosophy. Humdrum and workaday things can also give occasion to it. A wonderful example is the eulogy of love by Paul the Apostle:

> And though I have the gift of prophecy, and understand all mysteries, and all knowledge; and though I have all faith, so that I could remove mountains, and have not charity, I am nothing. (New Testament, 1. Corinthians 13, 2)

And how much astonishment and wonder may be inspired by a glimpse of heaven on a starlit night? So often we can only marvel at our own existence. And it is in this spirit, not from a biographical or emotional standpoint, but from a rational point of view, that we approach our question: what singles out near-death experiences from other psychic experiences?

2.3 Are Abnormal Psychic Sensations a Basis for Exceptional Reports?

The painting by Hieronymus Bosch is not the only item to be considered as reification of near-death experiences. Even reports in old texts are discussed because of possible similarities to near-death experiences. But is every report which contains similar or even the same experiences about a real near-death experience? As we shall see in the following, the term "near-death experience" comprises several different elements. These elements are the many and varied experiences. Each experience on its own is not unique to a state of near-death or clinical death, since such experiences also appear in other conditions of the brain. Those conditions include neurological and psychiatric diseases, conversion disorder (see table p. 53), and sometimes simply abnormal experiences by someone with a perfectly healthy mind, not to mention states of trance and possession and—last but not least—drug abuse. Before we go into detail regarding these examples, we can already identify a difficulty in deciding which of the historical reports are based on which cause or origin of the experiences.

There is another important point. In a strict interpretation of the words, near-death experiences are connected with clinical death, which means that they stand for experiences in a state of near death, immediately prior to the occurrence of death. Such a state of clinical death is potentially reversible by reanimation. This feature distinguishes clinical death from ultimate biological death. If reanimation fails, clinical death passes over to biological death. Cases in which clinical death is survived without severe physical or mental harm imply the availability of emergency service systems and intensive care medicine. Unfortunately, despite the high density of emergency facilities in modern industrial countries, only a few reanimations are actually successful. If we now consider the times when those old reports were drafted, reports today considered by some as descriptions of near-death experiences, we must take into consideration the number of people who might have survived a clinical death without major alteration. It is easy to understand that the number of such people might be extremely low, in view of the fact that there was only very crude medical support in those times.

On the one hand, people with extraordinary experiences or diseases always influence religious thinking in a society. Consider, for instance, what was known as *morbus sacer*, the holy disease, because there was no other explanation for epileptic seizures than possession of the afflicted person by some supernatural force. Or consider the case of Bernadette Soubirous, a girl from the city of Lourdes in the south of France, who had visions of the Virgin Mary in 1858, when she was 14 years old. At first, the Catholic church regarded these visions as an illness, but the ever increasing number of pilgrims seemed to demand a spiritual explanation. Even today—150 years after the visions of Bernadette—Lourdes still remains an important place of pilgrimage. As another example, in 1445, the shepherd Hermann Leicht from the village of Langheim in Germany had three visions of the baby Jesus. The result was the erection of a pilgrimage church Vierzehnheiligen (church of 14 saints) in the city of Bad Staffelstein. We can see here how people who have obviously had visual hallucinations have to a certain extent built up our religious views.

One would like to ask how far these phenomena have contributed to shaping religious convictions. From a historical point of view, two main hypotheses present themselves:

* A person with abnormal psychic sensations stands at the very beginning. The person's experiences spread round a certain community and become interpreted as the result of some supernatural influence.
* A society already has a theoretical concept on how to approach the "final questions". Then a person with abnormal psychic sensations comes along and these sensations are taken as a proof of the concept, or influence that concept.

The truth seems to be somewhere in the middle. Myths change over the years. Influence factors are cultural exchange with other societies and the perpetual impact of changes in one's own society. The latter is brought about by advance of (scientific) knowledge, but also by philosophical controversy with religious dogma. A compromise between the two statements could be that there is such a thing as a collective unconscious, as postulated by Swiss psychologist Carl Gustav Jung (born 1875, Kesswil; died 1961 Küsnacht). But the assumptions are vague, and transcultural similarities between religious beliefs may well have their origin rather in the above-mentioned Kantian questions (p. 7). After all, it is a basic desire of all peoples everywhere to answer such questions as these.

A fear of **end of life** and a **sense of life** affects everyone regardless of time and culture. Furthermore, other factors shaped early religious thinking: **cyclic elements** in seasons, **dualisms** such as heaven and earth, **fertility** of the earth

Fig. 2.4 Portrait of John Locke, lithograph, c.1840

(which gives birth to plants as though from nothing), the threat to life by **natural disasters,** and many more. This subsequently took on a life of its own and became incorporated in symbolism, rites, and concepts (see the discussion of darkness-light dualism or gematria in the following chapters).

With regard to near-death experiences, the low occurrence rate and miscellaneousness of experiences refute any argumentation in the style of Jung. On the contrary, modern psychoanalysis points to life events and individual development as shaping behaviour and perception rather than inborn or inherited beliefs. This reflects much older philosophical discourse, beginning with the Stoics and later evoked by English philosopher John Locke (born 1632, Wrington; died 1704, Oates) who declared the mind to be a *tabula rasa* at birth, whence only experience causes notions. Locke (1805) wrote as follows (Fig. 2.4):

> For to imprint any thing on the mind, without the mind's perceiving it, seems to me hardly intelligible. If therefore children and idiots have soul, have minds, with those impressions upon them, they must unavoidably perceive them, and necessarily know and assent to these truths; which since they do not, it is evident that there are no such impressions. (ch. I, § 5, p. 15)

Furthermore, he opined that simple ideas, "*the materials of all our knowledge, are suggested and furnished to the mind only by those two ways [...] sensation and reflection.*" (ch. II, § 2, p. 93)

Locke is the originator of empiricism and liberalism in philosophy. According to Russell (1969, p. 585), his political doctrines even became a basis for the American constitution. As Russell put it, liberalism "...*stood for religious toleration; it was Protestant, but of a latitudinarian rather than of a fanatical kind; it regarded the wars of religion as silly. It valued commerce and industry, and favoured the rising middle class rather than the monarchy and the aristocracy;...*" (p. 577).

Locke's doctrine that all our knowledge derives from experience—and so is a posteriori—is a counterpart to the philosophy of René Descartes (born 1596, La Haye en Touraine—today Descartes; died 1650 Stockholm) and the scholastics which had prevailed hitherto. Followers of Locke were the Irishman George Berkeley (born 1685, County Kilkenny; died 1753, Oxford) and the Scotsman David Hume (born 1711, Edinburgh; died 1776, ibidem) who consolidated empiricism. In fact, strict empiricism also raises some problems and leads to solipsism –negation of the real existence of all things which surround the self. So the debate went on, challenged by Immanuel Kant who proposed his "thing in itself" as a solution to the problem, but which also provoked counterarguments.

These irresolutions used the romantic movement in philosophy as an opportunity to revive the concept of dualism between body and soul and the possibility that man could get insights into the things that lie beyond. Near-death experiences played an important role in such argumentation, as we shall see in the following. But flowery and seductive philosophical concepts are not the best. Rather, one should prefer those which mirror life praxis. In this sense, empiricist philosophy gave way to a non-speculative, comparable, and calculable approach to nature in which scientific understanding is possible by observation and experimentation. But Kant's final questions remain open. It seems man has to face up to and live with imperfection. And near-death experiences do not contribute anything to resolve this impasse.

All in all, at first glance, it is indeed conceivable that some historic reports about near-death like states may have been influenced or provoked by people who really had survived a clinical death. On the other hand, one should make the following point:

Either a physical or psychic abnormality occurs very frequently or repeatedly, or many people share the same abnormality, or else it must be so impressive that the story goes 'round the world'. If these preconditions are not fulfilled, it is unlikely that the event would make any impression on myths or reports.

And this is most likely the case with regard to *clinical death*! The point is that the same experiences occur in other 'mental states' than clinical death—abnormal psychic sensations in psychic disorders, diseases, neurological diseases, drug intake or trance, and many more. In addition, religious concepts like the prominent role of light could have been shaped by such 'mental states', and the desire for a life after life is embedded in all religious concepts. When we glimpse old reports today, what we find is like a ball of wool, and it is not always possible to identify a thread that leads to the very beginning.

2.4 Arrived at a Marvellous Place—a Report from Plato

A remarkable account which is considered by most of the near-death literature to describe a near-death experience derives from the Greek philosopher Plato. This account can be found in his influential book "The Republic" (politeia). Plato was born about 428 or 427 BC in Athens or on the Greek island of Aegina and died in 348 or 347 BC. In 387 BC, he founded a philosophical school in Athens, in the park of Akademos. The name of the location was then attributed to the school: Academia, which is still used today to refer to the academic world in general. It remained in existence for some 900 years, until it was closed by the Byzantine emperor Justinian I because it taught gnostic ideas which competed with the then official religion of Christianity (Fig. 2.5).

"The Republic" belongs to the middle period of Plato's works. In the later chapters he deals with the evidence for the immortality of the soul and the fortunes of those who live a just life, fortunes which concern not only the world of the living, but also the afterlife. According to Plato, the tangibles one experiences during terrestrial life are in no way comparable in abundance and greatness with what happens after death. Plato refers to an unknown man by the name of Er, son of Armenios, who died in a war. Since this "son of Armenios" is not mentioned by any other ancient authors, it seems to be a metaphor implemented by Plato himself (according to von Kirchmann 1870). Indeed, metaphoric use of myth is typical in Plato's writing. He used it to prove his thesis, arguing that there exist eternal ideas which shape the world and even man himself. These ideas are universals from which all other conditions can be deduced. However, man generally has no insight into these universals, a situation that Plato describes in his allegory of the cave: man sees only shadows, not the light and the scenery outside the cave. But in certain circumstances, the soul may get some insight into the universals. Contradicting this, his follower Aristotle (384–322 BC) postulated that sensations

Fig. 2.5 The Academia school in Athens in the classical age. Steel engraving from 1895

are true, and that mistakes or falsehood are always the result of an incorrect interpretation of what our senses show us. So Aristotle preferred an empirical standpoint, where knowledge can be gained by induction. The idea was that one could approach the truth by noting similar and repeated findings and then conclude to fundamental principles by conjecture. The main difference is that universals are real and true in Plato's view, whereas they are not real in Aristotle's. This so-called universals controversy would be a key issue in philosophy from that time on, and it preceded the above-mentioned debate around empiricism in the modern age.

Returning to Plato's report on Er, son of Armenios, he recounts that, after 10 days, all the bodies on the battlefield but his were in a state of putrefaction. It is clear that he had only been in a state of apparent death, where life seems to be extinguished, but in reality is not. So Er woke up on day 12 and this is what he recalled:

> …when his soul went forth from his body, it journeyed along with many others until they arrived at a marvellous place, where there were two openings in the earth next to one another, and opposite again two others above them in the heaven. Judges were seated between them, and after they rendered judgment, they ordered the just to journey up through the heaven to the right, and fastened tokens on their front signifying the judgment rendered. The unjust they

Fig. 2.6   Plato, steel engraving from 1877, after a bust from the Naples National Archaeological Museum

ordered downward and to the left, with tokens on their back signifying all that they had done. (Plato 2006, p. 351)

The warrior's task was to inform mankind about the way things happen in the afterworld. He thus exercised a prophetic role: by referring to judgement in the afterworld, he demanded a moral life on Earth. Plato himself admits that his description of the afterworld contains something novel. Up until then, the most popular description of the afterworld was the one about Alcinous, ruler of the Phaiakians, as recounted in the Odyssey. The Phaiakians were Greek navigators from the island of Corfu. The Odyssey is assumed to have been written by the Greek author Homer (around 800 BC), but it is not clear whether Homer is just a fictitious name used to cover several different authors. Alcinous told Odysseus about his experiences in the underworld. But Plato's intention is to go beyond the scope of that story, as becomes clear when the first-person narrator says:

I will not tell you a tale like the tale Odysseus told Alcinous... (Plato 2006, p. 350)

This is contrasted with the landscape of Hades in the Odyssey. Here, it is described as a land or island where the dead themselves live. It is located at the "limits of the world" with the "deep-flowing Oceanus" and it is "shrouded in mist and cloud" (Homer 1962, book11). Plato's metaphor of judgement is different. The warrior who had been in a state of apparent death describes how souls are separated into good and bad. The first experience well-being in the incredible beauty of heaven, while the others can only weep after 1000 years walking in the underworld. Revenge is dealt out tenfold for every single injustice:

> …that is, once every hundred years, because this is the span of human life—in order that the punishment imposed might be ten times the crime. (Plato 2006, p. 352)

Later we read:

> After each company was there in the Meadow 7 days, they had to arise on the eighth day and begin a journey. In 4 days they arrived at a place where they saw, from above, a straight light stretched like a pillar through the whole heaven and earth, very like the rainbow but brighter and purer. They reached it after going forward a day's journey, and there, at the middle of the light, they saw the extremities of its bonds stretching from the heaven; for this light binds the heaven together, holding the whole revolving firmament like the under-girths of a warship. From its extremities is stretched the spindle of necessity, by means of which all the circles revolve. The shaft and hook are of adamant, but the whorl is a mixture of this and other substances. (Plato 2006, p. 353)

Plato described the bulges of a spindle which represent the eight spheres of heaven. First, the sphere of fixed stars, then Saturn and those planets which were known in Plato's time: Mercury, Venus, Mars, and Jupiter. The Moon and Sun were also considered to be planets. A similar view was put forward by Claudius Ptolemy (born c. 100, died c. 175 AD) 400 years after Plato, when he established his geocentric model. The column has a static function and stands for an *axis mundi* upon which the whole world is suspended. Apart from Plato's mention, the *axis mundi* is deeply rooted in many religious systems. It also appears in architectural symbolism as a demonstration of the power of worldly rulers and lords. This can be seen in the little brickstone cylinder in the Forum Romanum in Rome (Italy). It is a remnant of a temple dedicated to the gods of the underworld, a place regarded as the navel of the Roman capital (umbilicus urbis) and thus of the civilized world. It was also the meeting place of the underworld and the world of the living (Fig. 2.7).

Fig. 2.7   Umbilicus urbis Romae, Forum Romanum, Rome, 2012

Ziggurats and obelisks provide another example. They had many "functions" such as connecting the mundane world with heaven, or standing as an indestructible symbol of power, while the obelisk also symbolises fertility through its phallic resemblance. These early ideas are mirrored today in the ongoing competition to erect the highest structure in the world. Town planning under the dictatorships of Nazi Germany and the Soviet Union in the 1930s are a paramount example of how the respective dictators wanted to express their power by erecting the tallest structures in the world in Berlin (the "Great Hall") and Moscow (the "Palace of the Soviets"). We see here the common thread of primal ideas, with roots deep in history but perduring in our own time.

Returning to Plato, he goes on to explain how souls choose a certain way of life or become connected with certain living conditions:

After this again, the Interpreter laid the patterns of lives on the ground in front of them, more numerous by far than those who were present, and very various, for there were lives of all kinds of animals, and especially all kinds of human lives. There were tyrannies among them, some permanent, others destroyed in midcourse and ending at last in poverty and exile and beggary. There were also lives of men distinguished for beauty in respect to their form, others for bodily strength and athletic prowess, or for birth and the virtues of their parents; and there were lives undistinguished in these same respects, and so similarly for women. But no order of soul was present in them, because to choose a different kind of life is necessarily to become different in character. (Plato 2006, p. 355)

Exploiting the story of the son of Armenios, Plato described the transmigration of the soul—not only from man to man but also from man to animal. Now this motif, the transmigration of souls, was not an invention of Plato. It is deeply connected with hunting rituals in indigenous peoples, and leads back to a level of human society that long precedes the classical Greek period. Furthermore, in Plato's writings, the transmigration of souls could be regarded as a concept of development of the new from the old. Future and past, development and destruction are linked to each other in eternal cycles. The transmigration of the soul after judgement could be interpreted as a renewal of life, even though Plato makes no mention of this idea in his story of the warrior.

The idea of a soul leaving the body is one of several answers to the question of what happens after death. It is quite understandable that this might be one of the earliest religious views of mankind. Before people believed in the existence of a god or different gods, hunters in primitive societies believed in reincarnation. The soul would wander from man to animal and vice versa. Animals were an incarnation of supernatural forces and sometimes built a simultaneous existence with man (Eliade 1981, see also G. A. Wilken, quoted in: Freud 2000). And in Ancient Egypt, animals acquired the status of gods and fulfilled the same totemistic function (Beltz 1987). Thus, the idea of a soul that is independent of the body and which could leave the body under different circumstances is old and deeply rooted. It is Plato's hope that man will investigate the knowledge mentioned in his narrative and use it to distinguish a good way of life from a bad one. He also implies that an important precondition for such a pursuit is a preoccupation with philosophy.

At first glance, the subsidiary idea of judgement in Plato's narrative seems to be something of an oddity. In a similar way to the Last Judgement in Christianity, a judge assesses the life of souls. But Plato was a Greek, and his idea differs from the religion of his contemporaries. Indeed, according to Greek religion of the classical period, only someone who offended the gods would be punished, while the gods would not judge the failings of man. Plato's idea of judgement reveals an Egyptian influence that could also be found in Pythagorean philosophy (ca. 6th century BC) (von Kirchmann 1870). It is conjectured that Plato had travelled to Egypt, or even further, and had thus come into contact with certain oriental philosophies which propagated such ideas (Störig 1958).

The Pythagoreans were supporters of a doctrine put about by the Greek philosopher Pythagoras (born ca. 570 BC, Samos; died ca. 510 BC, Metapontum). He introduced the term "philosopher" in order to avoid being taken for a wise person ("sophos"), but instead to claim modestly that he was simply a person who loved wisdom. A philosophical school was founded around

530 BC in Croton (today Crotone, Italy) and continued to exist for several decades after the death of Pythagoras. The Pythagoreans developed mystical concepts about numbers and transmigration of the soul.

"All previous knowledge flows together in his [Plato's, a.a.] *philosophy—similar to a combustion point"*, as Störig (1958, p. 139, no English edn.) put it.

In this respect, the important role of light as a universal force and the concept of transmigration of the soul could also be traced back to the Pythagoreans:

> This life force is […] the world soul (mens universa) and it is located in the central fire. The sun as a glassy body is a distant echo of the central fire which borrows its light from it. Central fire and sun are not merely important for all parts of the world, but also for souls, because the central fire immediately spawns the souls of gods whereas sunlight spawns the souls of man. Thus, souls of man are also similar to light and parts of the divine world soul. (according to Döllinger, quoted in: Arnold 1870, p. 20, no English edn.)

With regard to transmigration of the soul, Pythagoreans taught

> …that a mass of souls hovers in the air. This mass of souls includes those who have not yet been in a human body and those who have once been in one but already left it. […] A deity forces souls into the dungeon of a body—similar to a grave—in order to atone for errors made in their previous extracorporeal state, according to the doctrine of the Pythagorean school. If they take good advantage of that time of penance and purification, they will be lifted again into the disembodied, higher state they previously possessed in space. But if they do not so use their time of purification, they will be severely punished and pushed into Tartarus [part of the underworld, a.a.]. Otherwise they will have to wander through several animal and human bodies in order to achieve purification. (according to Döllinger, quoted in: Arnold 1870, pp. 21–22)

It is also worth looking at the numerical symbolism in Plato's narrative. There are two clefts in the earth which symbolize divine dualism. A column of light appears on day 4. Four stands for twice the divine dualism and this reinforces the transcendence and the supernatural aspect of the appearance of light. The bodies of the dead warriors are putrid by day 10, just as injustice will be punished tenfold. The number ten stands for completion and perfection. Only if injustice has been punished tenfold will it be suitably avenged. Numbers associated with misfortune, such as eleven or thirteen, do not arise in Plato's story. Numerical symbolism also comes into Pythagorean philosophy (Biedermann 2004).

Fig. 2.8 Saints Paul and Anthony in the Desert, oil painting on wood attributed to Lucas van Leyden. (Born 1494, Leiden; died 1533, Leiden), vintage postcard

2.5 Caught up into Paradise—a Report from Paul the Apostle

The New Testament of the Bible consists of apostolic letters which "function directly as an indoctrination about duties of Christian belief and morality" (Weber 1916). Saul of Tarsus—later referred to simply as Paul—wrote more of these letters than any of the other apostles. He addressed most of them to Christian parishes, and fewer to peoples or single persons.

The biographical data on Paul are somewhat blurred. It seems quite possible that he actually met Jesus. His conversion experience as reported in the Acts of the Apostles seems to have taken place in 30 AD, at around the time Jesus was crucified, and we may suppose that Paul was murdered or sentenced to death in year 64 AD, when the Roman emperor Nero was persecuting Christians.

Paul made several missionary journeys. The second of these brought him to Corinth in Greece. In the first letter to the Corinthians, he explained issues relating to Christian conduct, or the Christian way of life. The second letter, dated by some exegetes to the year 57 AD, adopts a personal style. Paul was afraid that his doctrine might be in jeopardy in Corinth. For this second letter, he prepared his return to the Corinthian parish. It reveals his intimate feelings and mood, and this can be regarded as a proof of its authenticity. A passage in Chapter 12 of the second letter to the Corinthians seems eminently relevant to our topic:

> It is not expedient for me doubtless to glory. I will come to visions and revelations of the Lord.
> I knew a man in Christ above 14 years ago, (whether in the body, I cannot tell; or whether out of the body, I cannot tell: God knoweth;) such an one caught up to the third heaven.
> And I knew such a man, (whether in the body, or out of the body, I cannot tell: God knoweth;)
> How that he was caught up into paradise, and heard unspeakable words, which it is not lawful for a man to utter.

This quotation from the Bible is reminiscent of an out-of-body experience. The third heaven is another name for paradise. The number three symbolizes perfection and derives from a synthesis of divine dualism by the rule thesis plus antithesis equals synthesis (Biedermann 2004). In the line "*I knew a man in Christ*", Paul reflects on himself after his community with Christ (Weber 1916). He uses his out-of-body experience as a metaphor to emphasize his aim of finding a way back to the true belief—a message aimed at the Corinthians. He also idealizes delight; delight is so overwhelming that he even "forgets" his body. But is this kind of delight merely metaphor, or is it a real experience? Another event in Paul's life might support the latter suggestion.

His conversion to Christianity, expressed in the aphorism "from Saul to Paul", which suggests that he was making a volte-face, was the result of a dramatic situation during a journey to Damascus in today's Syria.

In Luke's Act of the Apostles (9,3–4) we read:

> And as he journeyed, he came near Damascus: and suddenly there shined round about him a light from heaven:
> And he fell to the earth, and heard a voice saying unto him…

Supported by his companions, he finally arrived at Damascus (Fig. 2.9):

> And he was three days without sight, and neither did eat nor drank. (9, 9)

This sequence, starting with aura symptoms and followed by loss of tone, is typical of epileptic seizure. The long-lasting symptoms could imply a semi-conscious state or status epilepticus, in which seizure persists over a long period. We might therefore conclude as follows. If Paul suffered from epilepsy

Fig. 2.9 Artist's impression of the conversion of Paul in an idealized form, with the appearance of Jesus. Painting by Anton Robert Leinweber. (Born 1845, Böhmisch Leipa/Česká Lípa; died 1921), vintage postcard

(Bradford 1974), his out-of-body experience would also fit into the range of symptoms, because the latter does occasionally occur in persons with epilepsy.

All in all, reports from Plato and Paul the Apostle describe experiences which resemble those of near-death, such as escape of the soul, a light or a beam or column of light, and meetings with other souls or the souls of decedents. At the same time, we must admit that religious and philosophical views always coloured these experiences, so they could rather be regarded as metaphors than the anamnesis of a state of consciousness. Reduction of these stories to a mere near-death experience falls short. But we also see that many experiences had already been connected with certain interpretations and views. This raises the following question: Had the pattern of near-death experiences already been influenced and shaped by our religion or cultural history, or vice versa?

In any case it must be acknowledged that, besides our (metaphoric) images and convictions about death, near-death, and the transition from life to death, quite different views also exist. In the ancient world, the transition from life to death was illustrated as walking through a door. Such images can be found on Roman sarcophagi and even in late antiquity, when the Christian religion prevailed (Fig. 2.10).

Fig. 2.10 Detail of a late antique sarcophagus from the Naples National Archaeological Museum, 2011

Fig. 2.11 Inner face of the cover of a sarcophagus from Paestum, exhibited in the local museum, 2011

The following example provides an extraordinary and intriguing illustration of the transition from life to death which has been controversial among scientists right up to the present day. It concerns the inner face of a Greek sarcophagus dated to the 5th century BC, discovered during archaeological excavations in Paestum (Italy) in 1968. Paestum is the modern name for the old Greek colony of Poseidonia. In a common interpretation of the scene, the image depicts a man diving from the Pillars of Heracles (see Fig. 2.11– right), hence passing beyond the known world into Oceanus, the river which encompasses the world. Oceanus represents the beginning of all things, so this means that, at the end of his life, the man returns to the origin of all life and the circuit is closed. If this interpretation were correct, it would anticipate the ideas of Plato. One may suppose that the illustration of the man diving into the water is based on the beliefs of the neighbouring people, the Etruscans, or that Greek settlers brought such beliefs to Italy, having been previously influenced by the Lycians (Pontrandolfo et al. 2011). The latter lived in Asia Minor in what is today part of Turkey. Once contact had been made, new beliefs would have spread around the ancient world through the regular exchanges between Greek colonies, scattered as they were around the Mediterranean coast.

Against the background of these two examples, we can already begin to break down the topic of near-death experiences. How much of such experiences is metaphorical, how much is secondary interpretation, and how much is a real, "neuropsychological" core symptom? This question will come up throughout the discussion in the following pages. When thinking about the door chiselled in marble or the painting of the man diving into the water, we may ask perhaps somewhat fastidiously: If such metaphors had continued to dominate over the centuries, would we have different near-death experiences today?

2.6 The Light was Such as I Cannot Describe to you—an Account by Gregory of Tours

The late ancient world was a time of change. The Roman empire had already disintegrated, and many successor states had been established. And yet the cultural influence of the ancient world was still palpable, albeit under the domination of the Christian religion. The transition from the 4th to the 5th century AD is regarded as a boundary beyond which ancient thinking was unable to permeate. According to the historian Rudolf Buchner (born 1908, Munich, died 1985, Wolfach), up until that time, a this-worldly view had dominated (Buchner 1956), that is, understanding the way of the world by dealing with the passions of people and the forces of history. God and the saints intervened in history to establish a new morality in this mortal world, according to the commandments of Christianity. But then, the question of the end of the world and the Last Judgment came to dominate. It seems likely that this transition was also induced by political changes. The safety and reliability of the previous empire had been replaced by instability. Governmental authority, administration, and jurisdiction were less influen-

Fig. 2.12   View of the city of Tours, steel engraving around 1860

tial, and daily life thus became more difficult to predict. As a consequence, the trend was to seek justice more and more in the afterworld. Besides, another direction of Christianity was also powerful; the so-called Arians. They were opposed to the dogma of the trinity. Despite the fact that they themselves consisted of several tendencies, they could certainly have been regarded as a formidable competitor for the "official" Roman or Catholic church. Some peoples such as the Lombards, for instance, were Arians. The Catholic church had already denounced them as heretics at the First Council of Nicaea in 325 AD. Oppression of other religious movements as well as conversion of the Jews under duress were the beginning of an altered understanding of Christianization. So this is the historical background for the reports we want to discuss here.

Gregorius Florentius (born 538/539 AD, Clermont-Ferrand, died 594, Tours), also known as Gregory of Tours, became bishop of Tours in 573. Tours is a city in the centre of France, which was in those days known as Gaul (Fig. 2.12). Among his distant relatives, a martyr named Vectius (Vettius) Epagatus is mentioned (Heinzelmann 2001). He was executed in Lyon (formerly Lugdunum) during the persecution of Christians in 177 AD, under the reign of Marcus Aurelius (born 121 AD, Rome, died 180, Vindobona/Vienna). A more famous victim of the same persecution was Blandina (born ca.150, died 177) (Eusebius 1937).

In Gregory's writings, which are known as the "History of the Franks", he reports in book VII/1 about the death and reincarnation of Salvius, later bishop of Albi, a city in southern France. Today he is better known as Saint Salvi (died in 584 AD). Gregory described how the monk Salvius once got sick and suffered from fever. Suddenly, the room in which he lay begun to shake and became suffused with light. Then he sighed out his soul. But next morning, the corpse started to move, and Salvius came back to consciousness. Three days later, he told the monks who had witnessed these events what he had seen in the afterworld:

> When my cell shook four days ago [...] and you saw me lying dead, I was raised up by two angels and carried to the highest pinnacle of heaven, until I seemed to have beneath my feet not only this squalid earth of ours, but the sun and the moon, the clouds and the stars. Then I was led through a gate which shone more brightly than our sunshine and so entered a building where all the floor gleamed with gold and silver. The light was such as I cannot describe to you, and the sense of space was quite beyond our experience. The place was filled with a throng of people who were neither man nor woman, a multitude stretching so far, this way and that, that it was not possible to see where it ended. The angels pushed a way for me through the crowd of people who stood in front of me, and so we came to a spot to which our gaze had been directed even while we were still a long way off. Over it hung a cloud more luminous than any light, and yet no sun was visible, no moon and no star: indeed, the cloud shone more brightly than any of these and had a natural brilliance of its own. A Voice came out of the cloud, as the voice of many waters. Sinner that I am, I was greeted with great deference by a number of beings, some dressed in priestly vestments and others in everyday dress: my guides told me that these were the martyrs and other holy men whom we honour here on earth and to whom we pray with great devotion. [...] (Gregory of Tours 1974, p. 387)

After his recovery Salvius was elected to the bishopric of Albi.

At first glance, the report contains near-death features, such as the emergence of a bright light, the sound of voices or other noises (rushing of water), and the appearance of decedents—the martyrs who had died for their beliefs. It is interesting that the whole account should contain so many references to Bible passages. In the above-mentioned paragraph, the voice which came out of the cloud could be found in the Bible:

> And I heard a voice from heaven, as the voice of many waters, and as the voice of a great thunder: and I heard the voice of harpers harping with their harps... (Revelation, 14,2)

There are several different ways to approach Gregory's story:

* Firstly, we might consider that Salvius really is recounting a near-death experience. We are told that he was severely ill and thus might mistakenly have been declared dead. However, in the narrow sense of the term "near-death experience", this was not such an experience, because the event cannot be considered to match up with a clinical death, that is, a heart and circulatory arrest. Here we have what we shall later discuss as multiple realizations of mental states. Similar experiences occur in different states of the brain—psychiatric or neurological—as well as in inflammatory diseases or clinical death, and in many other situations too.
* Secondly, the prospects of surviving a true clinical death, i.e., heart arrest, were considerably worse in those days, because successful reanimation depends on good intensive care units and emergency physicians—a standard of medicine that was quite unknown at the time of Salvius.

Comparisons with Bible passages reveal that the report was edited at some point to make it more authentic or to emphasize the eternal truths the monk had experienced. This could have been done by either Gregory or Salvius, since they were both very familiar with the Biblical texts. This account cannot therefore be considered merely as a description, since there can be little doubt that it also has a religious and political content. The possibility of glimpsing the afterworld demonstrates the extraordinary grace Salvius was given by God, while the relation to the Bible emphasizes his trustworthiness. All in all, in his later functions as a bishop, he proved to be a good representative of the true and correct religious orientation, at least as far as the Catholic church was concerned. In those days, people were very open to the tales of mystics, just as they are today to a very large extent. And thus—apart from the true cause of the experience—the report bears an important message.

2.7 A Fire that Reached from Earth to Heaven— Reports by Gregory the Great

There is one more Gregory we would like to discuss—Gregory the Great, later Pope Gregory I. Whereas the first Gregory lived in what is now France, the other was connected with the country now known as Italy, but their dates of birth and death do not differ so much. Indeed, they were more or less contemporaries. Gregory the Great (born around 540 AD, Rome, died 604, Rome) came from a family of civil servants. After reaching the level of city prefect, he decided to become a monk and later founded a monastery in which he became abbot. Even in his own day, Gregory was known as a strict and rigorous man. On one occasion, one of the monks died and when his corpse was being prepared for burial, three coins were found in his clothing. Although it was too late to punish him alive, since possession of money was strictly forbidden in the monastery, Gregory decided to bury him without ceremony in a pit, as shown in the illustration on p. 29.

Gregory's inauguration as bishop of Rome or pope is connected with several myths. It is reported that he first rejected the duty to which he had been elected, fleeing into the forest near Rome. However, a light beam, or according to another tradition, a pigeon surrounded by light, revealed his hiding place and he was brought back to Rome in a triumphal procession. At least, this is the version of German historian Ferdinand Gregorovius (born 1821, Neidenburg/Nidzica, died 1891, Munich), recounted in his History of the City of Rome in the Middle Ages, first edited in the years 1859–1872 (Gregorovius 2010). There were many hardships when Gregory the Great came to power. The Lombards threatened Rome, bringing with them the heretic religion of Arianism, and pestilence claimed many lives. In response, he officiated at a religious ceremony, whereupon a procession moved down from the Aventine Hill towards St. Peter's, crossing the river Tiber. But near the mausoleum of Hadrian, the archangel Michael appeared, drawing his sword, and this was the sign that the plague was over. So we learn that Gregory the Great's work was accompanied with divine power. Besides, today, Hadrian's mausoleum is also called the Castle of the Holy Angel.

Gregory the Great (Fig. 2.13) bequeathed a famous text known as "The Dialogues". He also established the concept of purgatorial fire in dogma and approved of force and violence in the conversion of people with other religions to Christianity. Gregory the Great published his work "The Dialogues" around 593 or 594 AD. The earliest Old English copies have been around since the end of the XIth century. The translation from Latin into Old English had been made two centuries earlier, at the end of 9th century, by Bishop Wærferths of Worcester, who was active from around 873 to 915 AD (Hecht in Wærferths of Worcester 1965).

In Chapter VII entitled "Departure of men's souls", which belongs to part 4 of "The Dialogues", Gregory reported as follows:

> ...I told you how venerable Bennet (as by relation of his own monks I learned) being far distant from the city of Capua, beheld the soul of Germanus (Bishop of the same place) at midnight to be carried to heaven in a fiery globe: who, seeing the soul as it was ascending up, beheld also. In the largeness of his own

Fig. 2.13 Gregory the Great punishes a monk who was convicted after his death of being greedy for money; wood carving around 1885, after a painting by Vasily Petrovich Vereshchagin. (Born 1835, Perm; died 1909, Saint Petersburg)

soul, within the compass of one sunbeam, the whole world as it were gathered together. (p. 188)

Another story deals with the faith of Probus, a bishop of Reati (today Rieti, a city in Lazio, central Italy). He was *"grievously sick and in great extremity of death"* (p. 192). Physicians had been called, but they were unable to save their patient's life. At one point, only a *"little young boy"* was present at Probus' bedside, and he…

> …suddenly saw certain men coming in to the man of God, apparelled in white stoles, whose faces were far more beautiful and bright than the whiteness of their garments: whereat being amazed and afraid, he began to cry out, and ask who they were: at which noise the Bishop also looking up, beheld them coming in and knew them, and thereupon comforted the little boy, bidding him not to cry, or be afraid, saying that they were the old martyrs St. Juvenal and St. Eleutherius that came to visit him: but he, not acquainted with any such strange visions, ran out at the doors as fast as he could, carrying news hereof both to his father and the physicians; who, going down in all haste, found the Bishop departed: for those Saints, whose sight the child could not endure, had carried his soul away in their company. (p. 192)

If we look closely at these descriptions, we once again detect features such as light in the form of a sunbeam, bright garments, or the appearance of persons in a scene suffused with light. We also find descriptions of the way the soul leaves the body.

In Chapter XXXI, "The death of Reparatus", Gregory recounts a tale told to him by a "friend and familiar acquaintance" (p. 214). The body of Reparatus had become stiff and everyone thought he was dead. But suddenly he returned to life and reported *that in the place where he was, he saw a great wood-pile made ready.* It was prepared for a man named Tiburtius, who was a priest but *"was much given to a dissolute and wanton life"* (p. 215). Reparatus describes a vision in which he saw Tiburtius' body burning on the wood-pile and:

> ...then another fire [...] was prepared, which was so high that it reached from earth to heaven. (p. 215)

Reparatus was unable to say for whom the additional fire was made, because he died while speaking. Tiburtius, although far from the place where Reparatus recounted this, was later found dead. This story is thus intended to demonstrate clairvoyance to the reader: during his apparent death, Reparatus was able to see into the future and witness things happening in distant places. Another chapter mentions the *"gift to speak with all tongues"* (XXVI, p. 209). One can also say that Reparatus' story reveals divine justice, since the "renegade" priest was punished by an early death. And we find a metaphorical link between the mundane and divine worlds, expressed by the fire which stretched from Earth to heaven. Furthermore, the similar concepts of light and fire illustrate a medium which, under certain circumstances, provides man with an occasional connection between the mundane and divine worlds. The same can be seen in Plato's "spindle", mentioned earlier.

Was Gregory the Great describing near-death experiences? Certainly not! All his reports were told by a witness, which implies someone who attended the dying person. These witnesses described what they saw and heard, but not what the dying person himself experienced. In clear contrast, near-death experiences are always told subjectively! But even here we find descriptive elements which resemble those of near-death experiences. Once again we come to the point that the very same mental experiences can be triggered by a variety of circumstances. Might it be that this obvious metaphoric use could be traced back to people who had survived a clinical death and whose experiences went "round the world", becoming well known to everybody? However, as mentioned above, the problem is that survival of clinical death would be very unlikely in those days. So we can argue that either the descriptions are based on the abnormal psychic experiences of certain people, or they could *only* be regarded as metaphors. Otherwise the truth may lie somewhere in the middle. But this discussion cannot be resolved, because a myth always has several different roots: observations of nature, religious conceptions of sense and the fate of human beings, and psychic abnormalities.

There was also a political motive in Gregory the Great's writings. It seems that he sent an edition of The Dialogues to Theodelinda (born around 570, died 628), the wife of the Lombard king (Petersen 1984). It is noteworthy that all the saints mentioned come from Rome or the surroundings of Rome. They are all Italians (Gregorovius 2010). Regarding the mystic events, Gregory the Great wanted to show his readers that heaven supported "his"—the Italian—church leaders. This implied that the church of Rome, "his" Catholic church, was the right one, because its dignitaries were directly favoured by God. What better arguments could he send to the Arian spouse of the Lombard king? At any rate, Theodelinda did indeed convert.

Against this background, some historians have argued that the mystical content of the stories was a sign that they had been written centuries after Gregory the Great. Indeed, some descriptions seem to contain gnostic concepts, such as a fragrance as a sign of a spiritual act. But he was a child of his times and gnostic concepts still accompanied "official" doctrine, with a certain amount of cross-fertilization. Such assessments are often applied when historians argue from our modern standpoint. Petersen (1984) described the attitude towards such wonders as follows:

> It had become customary by the end of the fourth century to attribute to a holy man miracles appropriate to his degree of sanctity. Consequently, those who had achieved a high degree of holiness [...] would tend to have attached to them stories of miracles comparable to those performed by the great prophetic figures of the Old Testament, by Jesus himself, and by the apostles. (p. 33)

2.8 Mani and European Spirituality

In early societies, the acquisition of nutrients and resistance to bad weather conditions played an important role in daily life and gained a transcendental importance. Furthermore, the changing seasons and climatic requirements for good crops, such as rain, flooding, and generally speaking, every aspect of the availability of water, are impossible to imagine one without the other!

When we celebrate Easter today, we do so at the time of the Passover festival, which was a celebration of springtime (Eliade 1981). In European and North-American culture, it includes Celtic as well as Germanic influences.

There are also reports from the Islamic country of Uzbekistan that women used to toss a dash of salt in the river Amudarya. This even happened in the 1960s when Uzbekistan was incorporated in the Soviet Union. Women worshipped the river goddess Anahida and hoped to get pregnant (Snessarew 1976). Incidentally, the goddess Anahida derives from the pre-Islamic period, but like many other beliefs, it survived the Islamic conquest and became integrated into the new religion. This example shows how water and fertility go hand in hand, and highlights the importance of water in a country situated in the hot continental climate of Central Asia and surrounded by semi-deserts.

In this vein, it is also important to stress the role of the sun, which is associated with light, growing, and coming-into-being. The Egyptian pharaoh Amenophis, also known as Akhenaten, worshipped Aten, the solar disc, as the only god. Moreover, Egyptians made their journey into the afterworld on a sun ship. Light, warmth, power—all these positive features are connected with the sun.

It was in the 3rd century AD that the Babylonian Mani or Manes (216 to 277 AD) made the dualism between light and darkness the main focus of his religion. As with Mohammed's religion 400 years later, Mani's belief system was also a syncretism, integrating Jewish and Christian elements. According to Mani, light itself is a component of man. Here is a fragment from the book entitled Kephalaia (meaning "discourses" in Greek), Chapter 38:

> So also this body! A mighty power lives here, even if it is small in stature. Nevertheless, sin dwells within, and the Old Man who is lodged in it. Certainly he is cruel, with great cunning; until the Light-Mind finds how to humble this body, and drive it (according) to his pleasure. [...] Now perhaps you know too that... the world is set firm, being ordered... the five sons of the Living Spirit in all its members... Sin took this body out from the land (?of Light)... constructed it in its members. It took its bodies from the five bodies of Darkness. It constructed the body. Yet the soul it took from the five shining Gods. It bound the soul in the five members of the body. It bound the nous in the bone; the Thinking in sinew; the Thought in veins; the Imagination in flesh; the Counsel in skin. (Welburn 1998, pp. 129–130)

Nous is a concept from pre-Socratic philosophy and stands for a principle which initiated the creation of the world but does not intervene in the course of events. We also find gematria in the text. The number five stands for a principle of order. According to the dogma that each human being mirrors cosmic order, this principle finds expression in the five parts of the human body (head, two arms, two legs), or in the five senses. Various aspects of light and darkness also appear in the text, further drawing attention to Mani's concept.

Light, considered as a part of what is good, is caught up in evil, embodied by matter. Only liberation of light, the good thing, can lead to ransom. "*The last particles of light will be brought together in a 'statue', which will ascend to heaven.*" (Eliade 1982, p. 392)

A fragment from the Fihrist, a book by the Muslim scholar Ibn al-Nadim (year of birth unknown; died 995) who lived most of his life in Baghdad, makes reference to Mani and tells us about the importance of light:

> Mani said, 'The origin of the world was composed of two substances, one of which was Light and the other Darkness. Each of them was separated from the other. Light is the great substance and the first, but not in quantity. It is the deity, the King of the Gardens of Light. It has five worlds: forbearance, knowledge, intelligence, the unperceivable, and discernment. It has also five other spiritual qualities, which are love, faith, fidelity, benevolence, and wisdom.' (Welburn 1998, p. 176)

Mani's doctrine is not completely innovative. The duality of light and darkness is borrowed from much older Iranian thinking. The main idea of Mani's doctrine of ransom is typical of Gnosis, and is even fashionable today. Light particles find their way back into the big picture, back into the world of light. Ransom means finding a way into the primordial state, back into a golden age.

But is this not a thought which also resonates today in our scepticism about technization and the manipulation of our everyday lives, given the dangers of war, terrorism, and climate cataclysms which threaten the whole of mankind? Light itself is simply a symbol of good things, of hope. We can see that the idea is still alive when we remember the tens of thousands of Leipzig citizens who performed the miracle of a peaceful revolution in East Germany in autumn 1989 by holding candles in their hands, a revolution which would have been impossible without the support of the church.

"*I am the light of the world: he that followeth me shall not walk in darkness, but shall have the light of life*", said Jesus, as it is written in the Gospel according to John (John 8, 12). And in the Quran, we may read: "*Allah is the light of the heavens and the earth.*" (24, 35)

In all cultures light is a sign of the transcendental or the divine. Equally, the dualism between light and darkness is deeply rooted in the religious life of mankind and has already become a part of our everyday culture. The main difference between Mani's doctrine and Christianity or Islam is the central importance of the duality of light and darkness. As mentioned above, this dualism goes back to old Iranian or Persian symbolism and the mythological fight between Ormuzd (light) and Ahriman (darkness). It resembles a struggle

between order and chaos, between good and evil, which has found its metaphor in light and darkness.

Furthermore, Mani's religious system explains nature in a pseudoscientific way by incorporating it into the religious system. The proof of the correctness of such "science" is actually circular. If one accepts religious doctrine without any doubt, then natural phenomena can be explained within the framework of the religious doctrine. This is then regarded as proving the truth of the religious dogma. Note in passing that Mani's doctrine, like other religions, follows Plato's form of cognition, starting from an eternal idea or dogma into which observations will be fitted, or indeed squashed when necessary. This approach was also zealously accomplished by Pythagorean and gnostic masters who dealt with number mysticism (gematria).

The Moon is a vessel into which light is poured, reported the theologian Ephrem the Syrian (born 306 AD, Nisibis/Nusaybin; died 373 AD Edessa/ Şanlıurfa) by citing Mani, although in a script opposing Manichaeism (Adam 1969, p. 15). Even though it attacks Mani's doctrine, it is important for philologists because the quotations and attributions allow us to compile original texts by Mani which had only survived in fragments or in quotations by other authors. From Ephrem and other sources, we know that, according to Mani, human souls go to the Moon. There they are committed to the arms of the father. The souls stay in the column of fame. This is the column of light which is full of purified souls (according to Flügel 1862). Once again, we find the positive associations of light already mentioned above in a similar context: the connection with the supernatural, regression to a (happy) primordial state, purity, and fame.

Mani also stated that man could have knowledge from former lives, an idea which has its origin in Pythagorean philosophy hundreds of years before. As for what happens when a man dies, according to Ibn al-Nadim's book Fihrist, Mani makes the following remark:

> When death comes to one of the Elect, Primal Man sends him a light shining deity in the form of the Wise Guide. With him are three deities, with whom there are the drinking vessel, clothing, headcloth, crown, and diadem of light. There accompanies them a virgin who resembles the soul of that member of the Elect. […] They (= three deities, a.a.) place the drinking vessel in his hand and mount up with him in the Column of Praise to the sphere of the moon, to Primal Man and Joyfulness, Mother of the Living, to where he at first was in the Gardens of Light. As for the body of the member of the Elect which is abandoned and cast down, the sun, the moon and the light shining deities abstract from it the forces which are water, fire, and ether (zephyr), and which

ascend to the sun, becoming divine. But the rest of the body, all of which is Darkness, is flung to the lower regions. (Welburn 1998, p. 151)

Mani's disciples proselytised and Mani's religious system subsequently became a major competitor for Christianity. As a consequence, followers of Mani's religion were persecuted by the Christians, and the last few strongholds of Manichaeism faded out in the 14th century in China.

Even though Mani's religion is not familiar to most people, a Manichaean tendency—as Eliade (1982) put it—is still an integral part of European spiritualism (p. 395). In this sense we can easily point to a multitude of current esoteric publications in the form of journals, books, or movies which tend to emphasize a positive role for light. Mani's is a gnostic doctrine which makes God himself, his aims, and his acts visible to those who gain pseudoscientific knowledge. However, this knowledge can only be obtained by people who belong to the elected circuit. It is not comprehensible to "outsiders". But is this not also typical of some of the views of near-death experiences which are currently propagated? Today, there are Internet networks, organisations, and self-help groups which claim to have knowledge about meeting with God or being adopted in heaven. Could these be the modern gnostics?

The various examples mentioned here show one thing particularly clearly: many of the phenomena occurring within a near-death experience involve spiritual views which became part of our mindscape, but also part of our wishes, eons ago. We should therefore ask whether old myths and lores are based on abnormal psychic experiences, whether they are only metaphorical, or whether they are a construct built up after they had become associated with abnormal experiences. None of these possibilities can be verified, and so they remain a matter of dispute. As is often the case, myths contain a core of truth in which reality, psychopathology, and fantasy meld together.

The most important insight from this discussion is that frequently mentioned near-death experiences such as perceptions of light, tunnels of light, leaving one's own body, and meeting with decedents have been a matter of religious dispute from time immemorial. And again we touch upon the question which is a guiding theme in this book: Which parts of the flowery narratives of near-death experiences are interpretation and worldview, and which could be regarded as a basic neuropsychological phenomenon subsequent to clinical death? And what about similar experiences occurring in completely different circumstances?

2.9 From a State of Apparent Death to Near-Death Experiences: New Wine into Old Wineskins?

Discussions of what we call today near-death or near-death experiences are older than many people assume. The terms may have changed, but much less the content! In particular, 19th century scientific publications about the state of apparent death anticipate much of today's controversy over near-death experiences. At the transition from the 18th to the 19th century, Europe was swept by a fear of apparent death. People were afraid of being buried alive with the consequence that they would eventually die under dreadful circumstances from suffocation or starvation in the coffin. It would not be an overstatement to describe the attention to the topic in those days as a form of hysteria.

It seems that publications early in the 18th century focussed on detection and avoidance of apparent death, while a new field of research developed in the 19th century: What did people who survived an apparent death experience whilst unconscious?

It would be surprising to find that public interest in "near-death" was limited to our own time, or even to the last two centuries. And indeed, there is an account by Democritus (about 460 BC, Abdera in Thrace; died about 370 BC) who discusses people who survived an apparent death in a lost fragment entitled *On the Things in Hades*. These remarks were delivered to posterity by a late ancient philosopher named Proclus (born 412 AD Xanthos?; died 485 AD Athens) who only mentioned Democritus' occupation, but gave no details (Die Vorsokratiker 2010, p. 313). In his Natural History, Pliny the Elder (born 23 AD, Comum; died 79 AD, Stabiae) wrote about a consul who had been declared dead but woke up during cremation and eventually died (Plinius 1994, book VII, chap.LIII).

There is also a story from the Middle Ages. The French king Louis IX or Saint Louis (born 1214, Poissy; died 1270, Carthage) suffered an apparent death when still young (Fig. 2.14). He was afflicted by a severe case of dysentery, and his physicians declared him dead. But then suddenly he came round and made a full recovery. He is said to have organized a crusade in recognition of this (Le Goff 1996). What is fiction and what is true can no longer be disentangled, but the intentions behind the story are clear: God is on the side of the "true" religion, of the "right" leaders. He brings the king back to life because he has not yet fulfilled his divine order—the crusade. Half a millennium later, apparent death had lost its divine function and become a matter of discussion for physicians. In the mid-18th century, the Danish-French anatomist Jacob B. Winslow (born 1669, Odense; died 1760, Paris) stated in

Fig. 2.14   Saint Louis, steel engraving from 1838

a publication that the assessment of signs of death is always uncertain, and that everybody was at risk of being buried alive. At almost the same time, the French physician Jean-Jacques Bruhier (born 1685, Beauvais; died 1756), working mainly in Paris, published a compendium of stories about the state of apparent death. Such influential publications and rumours about a Paris hospital in which dying people were enclosed in coffins before they were actually dead, as a way to get empty beds and thus avoid overcrowding of the wards, reinforced the fear of apparent death enormously (Necker-Curchod 1790). These fears even acquired a scientific name: taphophobia (Fig. 2.15).

Several mechanical devices were invented to avoid people being buried alive, allowing someone inside the coffin to establish contact with the outside world. Morgues were also introduced at the end of the 18th century. The physicians who instigated this were the Frenchman François Thiérry (born 1719; died 1793) and the German Johann Peter Frank (born 1745, Rodalben; died 1821, Vienna). Frank was the personal doctor of the Tsar Alexander I of Russia and and also of Napoleon. Another famous German physician who published on the subject of apparent death was Christoph Wilhelm Hufeland (born 1762, Langensalza; died 1836, Berlin) (Fig. 2.16).

Fig. 2.15 "The precipitated inhumation", painting by the Belgian painter Antoine Joseph Wiertz. (Born 1806, Dinant; died 1865, Ixelles), Antoine-Wiertz Museum Brussels

After studying medicine, he practised in Weimar and became the personal doctor of the Grand Duke of Weimar, Karl August (born 1757, Weimar; died 1828, Graditz). The influential poets Johann Wolfgang von Goethe (born 1749, Frankfurt/Main, died 1832, Weimar) and Friedrich Schiller (born 1759, Marbach; died 1805, Weimar) were among his patients. In 1798, he also became the personal doctor of the Prussian king Frederick William III (born 1770, Potsdam, died 1840, Berlin). He subsequently became director of the famous Charité hospital in Berlin. He was already interested in apparent death when working in Weimar. An early publication from 1791 addresses the issue of proof of apparent death (Hufeland 1791). Much more influential was his 1808 edition of "Der Scheintod" [apparent death] which became a widespread and very popular book. It can be regarded as an encyclopaedic work about the whole field of apparent death.

The opening of Europe's first morgue in Weimar (Germany) in 1792 was a result of Hufeland's efforts. According to Schwabe (1834), the idea of establishing morgues came from Johann Peter Frank who is appreciated for having established the foundations of hygiene and social medicine. In Paris, "La morgue" seems to precede the one in Weimar (Schwabe 1834, p. 10), but in contrast to the latter, the early Paris morgue was probably only a "storage place" without technical equipment. Equipment in the Weimar morgue com-

Fig. 2.16 Christoph Wilhelm Hufeland, lithograph around 1845. Note the artist's mistake in the first name, which should be "Christoph", not "Christian"

prised heating and ventilation systems and a room for a guard, with a window to observe the corpses in case they should reveal signs of life. Moreover, there was also "rescue apparatus": the limbs of each dead individual were connected to a bell by threads. The tiniest movement would thus have produced a noise, and apparent death could thereby be detected. The design drawing on page 40 (Fig. 2.17) shows technical equipment in the Weimar morgue of 1823. It shows the "alarm apparatus" (Weckerapparat). Down the middle are five threads holding thimbles, one to be put over each finger. Such a thimble is shown in "*Fig. 2*", while "*Fig. 3*" shows the alarm bell. The other figures show different profiles of a bed for the corpse.

Even literature dealt with apparent death. An example is the story entitled "The Premature Burial" by the American novelist Edgar Allan Poe (born 1809, Boston; died 1849, Baltimore). Many years later, the American novelist Samuel Langhorne Clemens, famous under the pen name Mark Twain (born

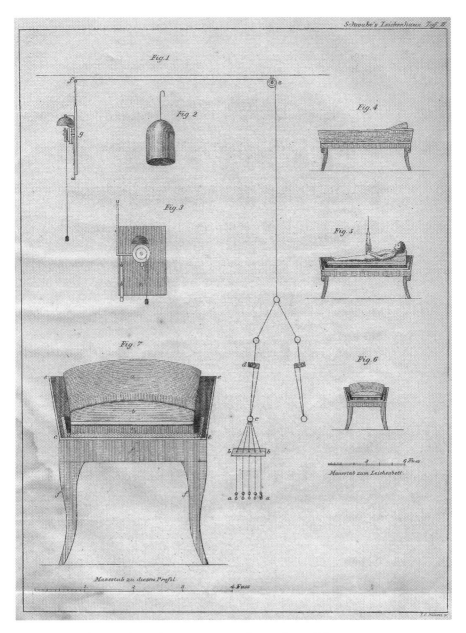

Fig. 2.17   Equipment from the morgue in Weimar, after 1823

1835, Florida, died 1910, Redding), visited a morgue in Paris during his travels through Europe in 1867. This was not the "early" morgue mentioned above, but the newly erected one of 1807. He noted that each dead person was placed on a block of stone, and people were allowed to look at them though a grating. Indeed, the morgue was open to the public. Twain noticed

Fig. 2.18   The Morgue in Paris, view from the outside, steel engraving around 1860

that many people only came for the excitement of this unbridled voyeurism (Fig. 2.18):

> Man and woman came, and some looked eagerly in and expressed their faces against the bars; others glanced carelessly at the body and turned away with a disappointed look—people, I thought, who live upon strong excitements and who attend the exhibitions of the Morgue regularly, just as other people go to see theatrical spectacles every night. (Twain 2010, p. 84)

It might be expected that early scientific work would deal especially with procedures for detecting apparent death, and indeed, as a side effect, modern thanatology was born. Signs of death that could be taken as certain were separated from those that remained uncertain. But in the early 19th century, more and more publications scrutinized what people experienced after an apparent death. Philosophical works had a considerable impact here, and discussions which began in those days anticipate modern debate about near-death experiences. In particular, the so-called romantic branch of philosophy mirrors today's spiritualist views. Romantic philosophy is said to have begun with the Swiss born and French speaking philosopher Jean-Jacques Rousseau (born 1712, Geneva; died 1778, Ermenonville) whose political reflections not only achieved success in France, but have also influenced the history of mind throughout Europe from that time on. But in the 19th century, Romantic philosophy flourished mainly in the German states. The Romantic movement itself stood for liberation of personality from social and moral conventions. The self became the centre of attention, and passions became all-important. Thus direct sensations and feelings were accorded considerable value (according to Russell 1969, p 651ff). Old Pythagorean concepts of transmigration of souls and the later gnostic dogma of emanation had their revival. The latter—emanation—stands for the gradual release of the spirit like a ladder from God to a world soul and finally to the soul of man, which contains part of the divine soul, a part of the big picture. These ancient concepts were combined with contemporary ideas of vital energy or animal magnetism, while the latter was also included in the concept of the (luminiferous) ether. In a nutshell, those concepts postulated energetic media that could function as spirit carriers, but were also used to explain phenomena like somnambulism or, in physics, the propagation of light.

The German-born philosopher Immanuel Hermann Fichte (born 1796, Jena; died 1879 Stuttgart), son of the philosopher Johann Gottlieb Fichte (born 1762, Rammenau; died 1814, Berlin), published a famous book about the human soul in 1856, writing: "*there are states of consciousness of the soul which appear independently from the organic apparatus of the body and its requirements and thus show a non-fixed consciousness*" (Fichte 1856, p. 398, § 166). He also declared the release of a "*primordial ability*" and a mental "*primitive state*" (p. 396, § 165). Here we find concepts of the continuity of mental states before birth and after death and a connection with the "golden age" of primeval times. So it is not surprising that Fichte pointed to reports of mentally ill people who became very clear in mind shortly before death, saying for example that "*the spirit loses inhibitions*" (p. 394), and hence also his positive attitude towards clairvoyance becomes understandable. To support his arguments, Fichte looked for demonstrations in the physics of the day, exploiting the concept of the luminiferous ether and declaring the spirit to be one of its various forms of energy. According to him, after death, man is just a spirit without hull. He also used this concept of the eternal soul or spirit to explain the origin of dreams, which are a "*sensation from outside which arouses the soul but induced by the brain*" (p. 405, § 169). To prove his arguments, Fichte illustrated with various examples, such as somnambulism, psychic sensations caused by the consumption of opium, cannabis, and monkshood (aconitum), and the case of a girl who fell over and became unconsciousness. All these cases shared "*an intensive and extensive enhancement of consciousness*" (p. 406, § 170).

In the romantic and idealist philosophy of the beginning of the 19th century, Plato's concept of eternal ideas which shape the mundane world was seized upon by philosophers like Friedrich Wilhelm Schelling (born 1775, Leonberg, died 1854, Bad Ragaz).

Fig. 2.19   Gustav Theodor Fechner, vintage photo

Another influential person was the German philosopher and physiologist Gustav Theodor Fechner (born 1801, Groß Särchen/Żarki Wielkie, died 1887 Leipzig) (Fig. 2.19). He postulated a world soul and a strict dualism of body and soul. The two exist in parallel, but also in a continual relationship of dependence. Such views influenced medical theories, like the theory of exogenous and endogenous nerve diseases due to the German neurologist Paul Julius Möbius (born 1853, Leipzig, died 1907, Leipzig). Möbius met Fechner in Leipzig and was heavily influenced by his ideas. These ideas particularly influenced physicians who dealt with psychology, which had not yet become a subject in its own right. In fact there was a considerable interplay between philosophy and medicine. Philosophers used examples from medicine to support their theses, while physicians used philosophical dogmas to explain their observations. But whatever ideas philosophers discussed about what underlies our world, the crucial question is: Could some people have the good fortune of being allowed to look "behind the scenes"?

The German physician and painter Carl Gustav Carus (born 1789, Leipzig; died 1869, Dresden) could be considered as a good example of someone who developed the interplay between medicine and philosophy. In 1846, he published an influential book entitled "Psyche. Zur Entwicklungsgeschichte der Seele" ("The Psyche. Historical Development of the Soul", no English translation). Carus introduced the term 'subconscious' into science. It was a conceptualisation which had its roots in ideas of the mathematician and philosopher Gottfried Wilhelm Leibniz (born 1646, Leipzig; died 1716, Hanover), and which became essential, half a century after Carus, for founding psychoanalytic concepts. Carus argued that the subconscious dominates the dream state *"which binds the dreamer closer to the general life of the world, which universalizes the person"*, explaining that *"this is why the dreamer is pervaded by all motions of the world."* Hence, even *"past and future, and time in general, interpenetrate and encounter one another."* In contrast, he related individuality, character, and freedom to waking consciousness. So Carus accepted and tried to justify divination through dreams (Carus 1930, p. 216).

An outstanding book with regard to near-death experiences is that of Franz Splittgerber (dates of birth unknown, died 1887?), a royal Prussian preacher of the Kolberg (Kołobrzeg) garrison. In 1866 he published a comprehensive and encyclopaedic book about sleep and death. More directly than Carus, he postulated an *"intense enhancement of the life of the soul during a dream [...] up to its highest peak, where dreams become a medium of divine revelation"* (p. 160).

In his book "Schlaf und Tod" ("Sleep and death", no English translation), Splittgerber reported the "symptoms" of people who had been in a state of apparent death. Among others he described noise, like the sound of bells, a noisy river, or the pitching of rods by helpers reported by a man who fell through the ice. According to him, others described *"overwhelming, sweet feelings"* or reported about a *"wonderful location"* where they *"saw several of their deceased relatives"*, or again: *"Everything is so nice and delightful there."* There are also cases with a *"prophetic talent"*, feelings of fear, visual phenomena like seeing a white garment, or a *"panoramic view"* of life, in which memories of life's events occur in fast motion. The author also described *"remote viewing"*, where people reported that they knew exactly what was happening in distant locations they had never been to before. Furthermore, *"doppelgänger"* reports were mentioned. This is where one sees oneself from distant positions, a term we could refer to today as out-of-body experiences or autoscopy. The author also recognized differences in the frequency of reports about such experiences. In women, the *"bond between body and soul [...] is looser, but in the same measure more elastic..."* This was supposed to explain why women more often suffered from *"syncope, catalepsy, magnetic sleep, and apparent death"* than men (p. 307). It is remarkable that Splittgerber revealed the multiplicity of triggers for such experiences, e.g., falling through the ice, *"death-like stiffness"*, *"frail farmer's boy with aching disease"*, and so on. Rather surprisingly, the author regards such discrepancies with a critical eye. He asks himself whether such experiences or reports should be considered as *"conceited creations of an excited and mentally ill fantasy"* or as *"impartial truths which give us an immediate insight into the things and processes of an otherworld."* He answers by saying that he *"would not exclude either, but would rather search for the truth in the middle,"* pointing out that experiences in the state of apparent death are too varied to approach the problem otherwise.

> The visions report on the otherworld and the elevating and harrowing events using exactly those pictures and symbols which the person also adopts when awake. In other words, exactly according to the religious view a person also has in ordinary life. (pp. 337–338)

If one strips off the *"subjective and fantastic garment"* of the experiences, then they *"harmonize with the basic religious views of God's word, that a retributive justice rules in the afterworld."* (p. 339)

Splittgerber pleads for a personal god and against pantheist theories, which were very popular in his time. These remarks and quotations highlight the importance of Splittgerber's book for comprehending the spiritual explanatory models of near-death experiences and the general debate on the subject in the mid-19th century. This includes the following problems, which we repeatedly encounter today in discussions about near-death experiences:

* the wide range of experiences,
* the fact that the same mental states can be realized by different triggers,
* the influence of personal religious and cultural background on the configuration of such experiences.

In Splittgerber's case, it is important to emphasize that he gave an impartial report. Of course, his account does mirror his own world view, but he scrutinizes many assumptions, especially the assumption that certain contents of these experiences, especially religious ones, are taken as premature evidence for the validity of the religion of the experiencer.

In 1861, the Dutch physician Alexander Willem Michiel van Hasselt (born 1814, Amsterdam; died 1902, The Hague) published a short compendium on death and apparent death which could be regarded as a precursor of modern thanatology. He defined apparent death as *"a state of an animal organism in which life has not yet completely ceased and in which life could become effective once more under favourable circumstances…"* (van Hasselt 1862, p. 84). He stressed that apparent death *"does not proclaim itself to the outside by sensorily perceptible appearances"* (p. 84). It is appreciable that, in this formulation, van Hasselt indirectly indicates the lack of methods to prove apparent death. Furthermore, he gave a general definition for *animal* life which included human beings—only 2 years after Charles Darwin's (born 1809, Shrewsbury;

died 1882, Downe) influential book "On the origin of species"! Van Hasselt discriminated different causes of apparent death like asphyxia, stroke, heart attack, intoxication, heat, and cold, and described symptoms such as headache, vision of sparks, visual and auditory hallucinations, and feelings of anxiety which could appear afterwards. Concerning cases in which people reported after apparent death about things that they might really have heard, he reasoned with the argument that, in apparent death, life is not completely extinguished, but reduced to a minimum. In addition, van Hasselt's book is important for a detailed discussion of reanimation, describing several different methods.

In many recent publications dealing with near-death experiences, reports by the Swiss geologist Albert Heim (born 1849, Zurich; died 1937, Zurich) from the end of the 19th century are taken as an example of extensive descriptions of near-death experiences. However, a closer look reveals that the condition for near-death experiences in the narrow sense of clinical death or at least apparent death is rarely fulfilled. Many of his accounts concern people in fear, in life-threatening situations, or people who had merely lost consciousness. He often described wounded soldiers, and in some detail, mountaineers who had suffered a serious fall but survived. He described the feelings of those people just at the moment they lost consciousness and reported, e.g., absence of pain, a clear mind, a review of one's own biographical history. *"Finally, the falling person often hears nice music and then falls into a wonderful blue sky with pink clouds"* (Heim 1892, p. 329). For someone who wanted to avoid either "physiological" or "philosophical" explanations, his observations were simply evidence that death occurs with great feelings of comfort. Death by falling is a "nice death" (p. 336).

Subsequently, in the medical scientific literature, more and more non-spiritual views began to appear and eventually predominate. In 1907, a popular scientific brochure was published about the secrets of death. This brochure was probably a revised and updated edition of a book from the mid-1850s, edited by the Leipzig-born physician and popular scientific writer Gottfried Wilhelm Becker (born 1778, Leipzig; died 1854, Leipzig). Contrary to its mystic title "Die Geheimnisse des Todes" ("Mysteries of death", no English edn.), the 1907 edition of the book is based on rational views. It quoted case histories from people who had survived a state of apparent death and reported experiences like reviewing their own lives, or *"revival of completely forgotten language abilities"* (p. 65). It is also argued that those experiences were not an example of *"increased mental abilities […] but rather their decrease."* Lack of oxygen is discussed in the brochure as the cause of *"irritations which are followed by a depression of the nerves"*. This *"depression"* would be accompanied by a *"mental excitement with delirium and hallucinations"* (pp. 66–67).

About 15 years later in 1923, internist, head, and owner of a hospital in Silesian Ober- Schreiberhau (Szklarska Poręba), Johannes Haedicke (biographical data unknown), published the book "Über Scheintod, Leben und Tod" ("On the state of apparent death, life, and death", no English edn.). He defines the term "apparent death" as follows: "*The state of apparent death is a state in which life as a whole is annihilated as a result of a change of existential conditions, but which can be recovered by renewal of the vital activity of those existential conditions*" (p. 288). This resembles the present definition of clinical death.

So all things considered, are today's quarrels about near-death experiences only new wine in old wineskins?

Whereas research has brought many new insights, especially from the medical and cultural points of view, the basic problems seem to remain unsolved. "Near-death experiences" is a term which—in its common use—is too broad, and thus often impedes structured research, making the issues too vague. A glance at earlier discussions shows that problems of designation and classification of the term "near-death experiences" have to be considered even today when research methods are more subtle than ever. It is also important to strip off prepossession by any kind of religious or world view. These appear only in a secondary way, as everybody has his own personal world view.

3

Problems and Contradictions

3.1 Non-Specificity of Experiences

Amazingly large numbers of people claim to have had near-death experiences, according to the literature. One study shows (Schmied-Knittel 2006) that four per cent of all Germans report near-death experiences they have had during their lives. Studies from the USA show similar or even higher frequencies. Related to Germany with a population of 82 million inhabitants, this would mean that approximately three and a half million people have had near-death experiences.

But a near-death experience implies being close to death, i.e., in a state *near death*, at the cutting edge between life and death. So a near-death experience, in the proper meaning of the word, is connected with surviving what is known as clinical death—an emergency state, a life-threatening condition in which the organs begin to stop functioning, but still without irreversible damage to those organs, so that resuscitation could nevertheless succeed. Clinical death is defined by complete circulatory arrest, and hence a lack of pulse, and breathing arrest, but still reversible by means of reanimation. If the latter is not successful, damage to the organs increases and leads to brain death, because the brain is one of the most sensitive organs and it can only resist for a few minutes, and finally biological death, the end of all organ and cell function.

Circulatory and breathing arrest lead to a sudden decrease in the supply of nutrients and oxygen to the organs, where glucose plays a prominent role. It is thus hard to understand why some authors categorically neglect the fact that near-death experiences have something to do with a lack of oxygen or circulatory arrest. At the very least, these are the mediators of severe brain malfunction in *clinical death*! The period of time over which reanimation could be successful differs from organ to organ. An important limiting factor for a successful reanimation or a chance of surviving clinical death with no mental damage is the resistance of ganglion cells in the brain. At most, they can survive eight minutes. After successful reanimation—when oxygen and glucose supply functions again—they still need a certain period of convalescence, to reestablish themselves as it were. So normal function comes not immediately, but step by step. These facts are important for models explaining near-death experiences, and they are also important when we raise the question of the exact point of time when near-death experiences develop. Other organs can overcome a clinical death that lasts much longer: the heart 30 min, the lungs 60 min, the kidneys even 2 h. In clinical death in children or people with hypothermia—i.e., breaking through the ice—the survival time can be considerably longer.

It is interesting to ask what the high number of near-death experiences tells us. To get an idea, we can look at the reanimation register. In Germany, the number and outcome of each reanimation is collected anonymously by an initiative called the German Resuscitation Registry GRR. Calculated for a city of 100,000 inhabitants, 55 to 66 patients are reanimated each year. This corresponds to a percentage of 0.07. Extrapolated to all inhabitants in Germany, this implies that a total number of 56,000 people are reanimated each year. But an important precondition for reporting a near-death experience is success of reanimation. First of all, the person has to survive and get back to life, and, secondly, they must be in a psychic and physical condition that allows them to report their experiences. When we talk about statistics, a successful reanimation also includes survival with severe brain damage or a care-dependent state. This dramatically reduces the number of people who might subsequently report their near-death experiences.

Unfortunately, less than two percent (exactly 1.8 %) of all resuscitations are successful. That means, in Germany, we have no more than 1,000 people per year who are snatched back from clinical death and return to life. But again, how many of them are lucky enough to get back to a *healthy* life with normal mental functions, sufficient to be able to report their experiences, if they even had such an experience, and there are only a few, as we shall discuss shortly.

Even if we extrapolate the number of reanimations to the years in an average life—the above-mentioned survey is a longitudinal section, whereas the numbers in the resuscitation register derive from a cross-section—even in this case, the number of successfully reanimated individuals is much smaller than the number of people who claim to have had near-death experiences in their lives. Looking at this another way, it is clear that many people who have never been clinically dead in their life report experiences they define as *near-death* experiences! So there must be something wrong here! To come straight to the point, proponents of a spiritual view have long been glossing over these discrepancies. In the typical literature, we find a ready failure to distinguish:

* real near-death experiences, which means, in the narrower sense, experiences connected to survival of a clinical death,
* hallucinatory visions of dying people,
* fear-death experiences, which happen in life-threatening situations but have nothing to do with a clinical death, and

* experiences with the same content are sometimes reported after a clinical death, but which occur completely independently of the above-mentioned situations.

It should also be noted that not all reanimated patients have near-death experiences, as Martens (1994) showed in a systematic survey of patients who survived a cardiac arrest. The author concluded that near-death experiences are very rare. Newer investigations have revealed that the typical and popular sequence of experiences as propagated by Moody has no constancy, and that in fact the various experiences differ in their frequency. The most prominent experiences seem to be feelings of warmth and peace, according to the author, with a frequency of about 60 to 89 % among those who report near-death experiences. Visions of light occur in 77 %, out-of-body experiences in 24 to 61 % (Moody 1975; van Lommel et al. 2001), and 15 % report scenic or film clip recollections of their past (Fenwick and Fenwick 1997).

In our own study conducted in Uzbekistan (Engmann and Turaeva 2013), we found a similar diversification. We chose a central Asian country because of its unique circumstances with regards to near-death experiences. It is a country in which the Muslim religion is widespread, but which was also under atheist influence for about 80 years when it was part of the Soviet Union. Furthermore, many local traditions still play an important role in religion. We expected a population that was poorly informed about near-death experiences and would therefore provide less "prejudiced" reports, in contrast to reports from "Western" people who have mostly already thought about the problem of near-death experiences before their personal experience of it, or perhaps received "orientation" after their experience, from the internet or the press. Furthermore, data on near-death experiences are rare from the Islamic world, and may offer interesting contrasts to Christian views of it.

As in many near-death studies, owing to the low sample size, the significance of this study could always be discussed controversially. We were pleased to find 13 interviewees, formerly patients of an Uzbek hospital who had undergone operations or suffered from severe diseases. Most of them had been reanimated. In fact, 10 of the 13 had been reanimated and 3 had not. But not all of the reanimated patients had had near-death experiences. On the other hand, 2 of the 3 who had not been reanimated had had such "near-death" experiences. Even in that tiny sample we see that the same experiences are not necessarily connected with a clinical death, implying that the same experiences can be evoked by different causes.

The most commonly reported phenomena were acoustic experiences ("voices" and "noise" together), light, feelings, and out-of-body experiences

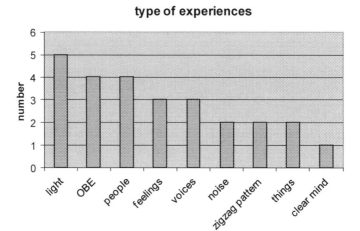

Fig. 3.1 Frequency of mention of different "near-death experiences" in Uzbek in-patients. "Feelings" include softness and smoothness or "tissue-like" feelings. OBE stands for out-of-body experiences. (Modified according to Engmann and Turaeva 2013, p. 6)

(OBEs) (Fig. 3.1). It is surprising that the sample did not include tunnel phenomena, which are generally viewed as the most prominent and impressive form of near-death experience, and nor did film-clip experiences occur in the sample. These findings go along with the above-mentioned distribution of the frequency of experiences. But we should always bear in mind the low sample size and avoid hasty conclusions! Even so, it seems that the pieces of the jigsaw fit together. We shall return to the above study when we come to discuss the presumed peculiarities of worldwide reports.

Athappilly et al. (1975) also called the tunnel phenomenon into question as a basic near-death experience—and quite rightly, as we have just seen. This suggests that there is no cascade of experiences which would occur regularly and completely in clinical death!

There is also the fact that the phenomenon of out-of-body or ecsomatic experiences even occurs in 8 to 10 % of the healthy population and in 50 % of those who abuse drugs (Kasten and Geier 2009). With regard to the healthy population, Green (1968) observed:

> The psychological circumstances which precede ecsomatic experiences, particularly 'single' ecsomatic experiences, are frequently characterized by the presence of some identifiable form of stress. In the majority of 'single' cases this stress is associated with a physical trauma of some kind, such as illness or accident. However, in one out of every four 'single' cases the stress is of purely psychological origin. (p. 25)

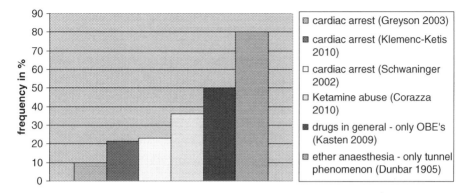

Fig. 3.2 "Near-death experiences" brought about by different triggers

(Incidentally, "single cases" refers to the author's finding that, in most of his subjects, out-of-body experiences usually occurred once and—less frequently—twice during the subject's lifetime.)

Taking this into consideration, the same or even a greater proportion of out-of-body phenomena occur in people consuming drugs than in people who have survived a clinical death. Does this not mean that out-of-body experiences would be better regarded as a symptom of drug consumption than as a characteristic of a survived clinical death? Despite this, for some people, out-of-body experiences are still a central phenomenon in the discussion of near-death experiences, because they are regarded as a proof of the dualism of mind and body, or a proof of the existence of a soul that is independent of the body.

The figure (Fig. 3.2) shows the occurrence of "near-death experiences" in different circumstances according to some selected studies. We observe that circumstances which have nothing to do with clinical death or cardiac arrest prevail with regard to generation of "near-death experiences." The two bars on the far right only consider out-of-body experiences and tunnel phenomena, respectively. Even here, the frequencies are quite high.

When we wish to appraise how often near-death experiences occur in different circumstances, this always reveals a great deal of uncertainty, and this uncertainty has its origin in the term "near-death experience" itself. Because no clear definition exists to tell us which experiences *must* be included in the term, one can always arbitrarily match experiences which—for instance in the case of drugs or anaesthetics—had formerly been referred to as hallucinations to the term "near-death experience". Here there is a kind of rededication. How could a drug consumer have a true "near-death experience" if he had been as fit as a fiddle whilst smoking his joint? The consequence of these problems is bias in all such investigations.

The best thing to do would be to avoid such a vague term as "near-death experience". Instead, one can report about the individual experiences which occur in clinical death or under drug consumption. They may be similar or even identical, as happens with many mental functions that can be brought about by different triggers. This whole problem is not new. It had already come into being shortly after publication of Moody's influential book. Hence, in 1981 Gabbard et al. concluded that no characteristics exist which exclusively define near-death experiences after clinical death. So criticism of the term "near-death experiences" was already a matter of discussion in this thirty year old publication (Gabbard et al. 1981).

And what degree of influence can be attributed to suggestive elements in questions and questionnaires about near-death experiences? It seems that many criticisms and doubts have never given this sufficient consideration over the last thirty years!

All in all, we may wonder whether it is justifiable to use the term "near-death experiences" at all. Does the locution "near-death experiences" encompass any well-defined entity? Or could "near-death experiences" be regarded as a term that was a useful auxiliary construction in the early phase of near-death research, but which is now—from a strictly scientific point of view—unsustainable?

3.2 Near-Death Experiences in Different Cultures

Most reports of near-death experiences derive from the so-called Western world, that is, Europe and North America in particular. Culture in these regions of the world is mostly shaped by Christian and Mosaic values, although we have already discussed gnostic and Manichaean influences which are deeply interwoven with both religions. If we want to discuss which of the experiences in near death are shaped by religious and cultural convictions of the relevant person and which are not, we have to look for reports from outside Western-dominated cultures. If we can figure out which experiences are not shaped by such convictions, it might reveal certain "neuropsychological core symptoms" showing affection of the brain in near death.

But there are two points to consider. First of all, the number of reports from "non-Western" cultures is rather small. And, secondly, all cultures are involved in exchange and are thus exposed to the impact of the dominating Western culture. These factors make it difficult to draw conclusions.

According to Ring (1980), near-death experiences occur in all cultures and religions. Some authors such as Kellehear (1993) or Feng (1992) have emphasized the predominant influence of cultural and religious factors on the way people describe experiences. This is supported by a publication from Augustine (2006), who cited investigations by Murphy with Thai people who had reported near-death experiences. The tunnel phenomenon, euphoria, and perception of light do not occur. But it is especially the last two of these experiences, positive emotions and light, that dominate narratives about near-death experiences in European and North-American literature. On the other hand, the experiences of Thai people include hallucinatory and emotional elements, such as perception of landscapes and unpleasant (!) feelings, or a ride into hell.

A few years ago, Belanti et al. (2008) compiled a literature survey. It contains many reports from South-East Asia, China, and Africa, but also cases from Hawaii, from native North Americans, and from the "Western" culture.

It is clear that most reports mention meetings with decedents or their ghosts or souls, then out-of-body experiences follow with a range of different frequencies, but tunnel phenomena were not reported in all cultures. The latter could be found in reports from China, Thailand, and the Netherlands, but—amazingly—not in Germany or Africa. The key to the problem is surely the sample size. For Africa, only 15 cases were used, whereas it is a whole continent with manifold cultures and religions. From Thailand, there were only 10 reports. Therefore it is difficult to draw quick conclusions. But descriptions nevertheless reveal some cultural peculiarities.

> [...] experiences in Thailand and India do not report visions of 'paradise', but instead have experiences that include visions of religious figures and the experience of Karma and judgment. (Belanti et al. 2008, p. 129)

Another point is interesting. All the Indian reports, making a total of 45, contained experiences like "residual marks on body" (Belanti et al. 2008, p. 125). The affected people told the interviewers that such signs were remnants from events that happened in the hereafter. Some remembered having had stigma and burns on their bodies. People from Northern India reported that they were knocked down palpably or by means of a device, a trident for instance, at the very moment when they resisted to leave the afterworld. In contrast, the people from Southern India believed that everyone that comes back from the afterworld into the mundane world usually gets a shibboleth. Unfortunately, these reports from Pasricha (1993 and 1995) were recorded long after the presumed "near-death" situation occurred. So many questions remain unsolved:

* Are the "signs" that people reported real scars connected with the circumstances of clinical death, for example, scars remaining from an operation or from an accident which the people had interpreted in a religious sense?

* Or are they older signs or abnormalities that were reframed after the "near-death" event, may be in a delusional way?

Another problem with such reports is that we cannot check whether the affected people really have survived a clinical death. No attempt was made to equate the stories with a medical report or an old hospital chart. On the one hand, these problems diminish the validity of the results, but on the other they once again highlight the already mentioned problem that features of near-death experiences can occur under many other circumstances than clinical death alone. This is an additional argument for a critical view of the term "near-death experience".

Meeting with Yama often occurs in Indian near-death reports. The Vedas, a collection of Hindu religious texts, tell us about Yama. The name Yama stands for sibling. In religious history, Yama initially had a twin sister called Yamun or Yam. She sacrificed herself by taking her own life. Yama thereby became king and vanquished death in the realm of the Blessed. He subsequently became the ruler of the netherworld. The Yama character also contains old Persian roots relating to the sun god Vivasvant (according to Zimmer 1980; Mylius 1978).

Yama is regarded as the first man in the oldest Indian mythology. He showed the way into the afterworld to all following men. Yama is located in third heaven. At this point, we once again encounter the kind of numerical symbolism that we also found in the above-quoted passage from the Bible: third heaven is the location of paradise.

Incidentally, via the Pythagoreans and Plato, who were almost certainly influenced by "Eastern" religions, numerical symbolism perdured in Gnosis and influenced Christianity, and even our thinking today. Take the example that all good things come in threes. Even many medical classifications use a trinity of symptoms, as demonstrated by Parkinson's disease: rigor (stiffness of muscles), tremor (jittering), and akinesia (lack of motion). And many people refuse to marry on a Friday 13th. In this sense, ancient India still influences the Western world today.

The third heaven is the heaven in which the gods of light (!) reside. It is home to dead heroes, the pious, and the generous. It is the location of an "eternal feast". By burning the dead, man is sacrificed to the gods and can thus be renewed and become godlike in a new body.

In Brahmanism, doomsdays are a key focus of the doctrine. The most important task of a human being is to do penance for the development of all evil in the world, and the highest fortune is to be completely rid of the sinful life. An entirely good man will be reborn as a god, while bad people transform into animals, or in the worst case, into plants. Plants are regarded as completely helpless—they can neither escape from danger nor struggle. In this vein, for instance, people who had stolen dishes came back to the world as bloodthirsty animals (Henne-Am Rhyn 1881).

In India such beliefs are still widespread, and even in our Western world, such ideas provide a fertile ground for esoteric publications and so-called "new-age" therapists. But such esoteric beliefs do not build up a closed circuit of evidence in which one finding explains the other. Quite the contrary. Various ideas from different cultural and religious systems get mixed up, and convictions are taken arbitrarily so that they somehow fit together. In our culture, myths are taken from the Jewish and Christian traditions, not to mention the Indian ones too. On the other hand, the lack of Islamic influence in Western esoteric concepts is rather surprising. It illustrates how concepts are influenced by attitudes towards other societies and religions, and also by politics.

Ancillary, mercantile thinking plays an important role too. We can think about parapsychological therapists who offer so-called past life regression—for money, of course! Solvent clients learn that their problems and worries today derive from a former time when they lived as another person, even in another country. This concept clear carries with it the idea of transmigration of the soul. It is a good thing that such therapies do not have juridical consequences, otherwise someone who was told in expensive therapeutic sessions that he used to live as a vintner in Tuscany in Roman times could perhaps claim a house and vineyard in Italy.

The possibility of punishment after death also exists in Buddhism, which was founded by a king's son named Siddhartha (born 623, died 543 BC) from the family of Gautama. Later he was called the awakened or wise. This is the meaning of the word "Buddha". Here, Yama is a judge in hell, where the time of punishment is restricted and can be followed by transmigration of the soul.

There are only a few reports or investigations about near-death experiences (NDEs) from the Muslim world. Kreps (2009) pointed out the strikingly low number of reports about near-death experiences in Islamic countries, unlike in the West. He concluded that

> NDEs are specifically designed for people who need them, and the need in certain communities may not be as great as because of the persistence of traditional faith in an afterlife and a creator. (p. 67)

In contrast to this position, Fracasso, Aleyasin, and Young (2010) reported about 19 Iranian Muslims who had had near-death experiences. The authors saw many similarities between their near-death experiences and those of Westerners with regard to content and after effects. They also disagreed with the position that near-death experiences are a rare phenomenon in the Muslim world. Nahm and Nicolay (2010) also argue along these lines. On the other hand, in these reports, the definition of the term "near-death experience" used

Fig. 3.3   Detail of an ornamental wooden door in Khiva. (photo 2010)

by the authors was insufficient. Such near-death reports are not only from people who had really been clinically dead, but also (and even mainly!) from people who had suffered a wide variety of triggers other than clinical death, such as "accidents" or "emotional shocks" (see, e.g., Fracasso et al. 2010).

It is clear that many things recounted by people with near-death experiences are an excerpt from their manifold religious ideas. The desire to meet dead relatives, positive associations with light, and transmigration of souls as a counterpart to out-of-body experiences are perfectly coherent with different cultures. Small wonder perhaps that many reports of near-death experiences resemble each other. Or could there be a neurophysiological reason for that resemblance?

In our survey (Engmann and Turaeva 2013), we found a particularly interesting report:

An Uzbek woman of Muslim beliefs in her mid-thirties explained that when she gave birth to twins something "broke" and she had severe abdominal pain. She was brought to hospital. In the course of the stay her situation got worse and she had to be reanimated. One of her recollections was that she flew to a door which was, according to her, impossible to describe, but with ornaments all over it; not like in Khiva. *"You haven't seen this before!"*, she exclaimed. What was the colour of the door? *"Light!"* and then: *"yellow"*. *"It must be paradise."* (p. 3)

Khiva is historical city in Uzbekistan, located on the Silk Road, and it is famous for its architecture. In fact, it is a UNESCO world heritage site. The historical buildings comprise various details such as a wide range of elaborately carved wooden doors. And even today carvers produce many items in the traditional style. The photo in Fig. 3.3 shows an example from Khiva.

In this report, we can argue that it was influenced by the world which surrounds the patient in everyday life—the local architecture! Is the door the patient had seen just a memory that was activated during the circumstances of reanimation? Or does it reveal a neuropsychological sign: did she experience a zigzag pattern which, together with light (!), could be regarded as an occipital lobe malfunction that was interpreted as ornaments? We shall return to this idea later.

3.3 Cluster Structure of the Term Near-Death Experience

Many terms in our everyday language can be described by associating various numbers of single criteria. Consider, for example, how we might explain the word "candle" to an imaginary foreign visitor with a poor understanding of the English language. A cylindrical, elongated object made of paraffin with a wick that burns. If we take each criterion on its own, our foreigner is unlikely to guess "candle", i.e., if we only tell him *one* of the words, like "cylindrical" or "elongated". But the more criteria we get across, the more exactly we describe the thing and the more likely it is to be understood.

So what have candles got to do with our discussion about near-death experiences? When a term consists of several criteria or minor terms, this is called a *cluster* structure. Such cluster terms are not only to be found in daily language, but also in medical terminology, including the context of near-death experiences. A crucial question is whether minor terms are specific for generic terms or not. It seems that they are not, as our "candle" example shows. Some philosophers also argue that there are exceptions. Going back to our example, candles are usually made from paraffin, but paraffin is not specific to candles. On the one hand, a candle could also be made from beeswax. On the other, paraffin is connected with other items, such as crayons, phlegmatizing agents, surf wax, and so forth. In this way, we can try to analyze every term in our language. We soon find that terms are rarely specific. Furthermore, we can even analyze the minor terms into more minor terms, since each minor term also consists of minor terms. For example, a wick is a thread and it is flammable, and a thread is long and thin, and so we go on. The task is never-ending!

The term "near-death experience" also has such a cluster structure. It consists of different minor terms, namely the different experiences such as noise, the feeling of being in a tunnel, visions of light, lifelike recollections of dead relatives or friends, film-clip scenes from one's own life, visions of geometric structures, out-of-body experiences, and so on (see Engmann 2011c). As we shall see in the following, none of the minor terms or experiences is essential for defining the term near-death experiences, and nor are they an essential attribute in the sense that, without them, the term "near-death experience" would be irrelevant. For instance, experiences of unpleasant noise could also occur in people who suffer from schizophrenia. In this case, such noises are called acoasms. Visions of light or visions of geometrical patterns also occur in the case of strokes or bleeding in the occipital lobe of the brain, or just in the case of a migraine. All in all, a single minor term or experience does not automatically reveal its cause, so we need further information.

A central question is: how many minor terms are necessary to define a term or a diagnosis sufficiently well and clearly? Or in this case, how many constitutive experiences define the term "near-death experience"? Moreover, are there minor terms or experiences which are *essential* for near-death experiences? In other words, are there experiences which, if they did not exist, we would have to scrap the term "near-death experiences"? Anticipating the following discussion, we can answer the latter question in the negative—such essential experiences do not exist!

A cluster structure of terms is a frequent phenomenon in the medical sciences. We may even think about the identifying symptoms of diseases which are preferentially grouped in threes, as already mentioned above. Let us return to our example of the trinity of guiding symptoms for Parkinson's. If we take each on its own, we observe that it could also be found in other diseases. Moreover, many more symptoms could be associated with Parkinson's. In conclusion, terms with a cluster structure can never be defined accurately and without contradictions. They always contain an element of uncertainty, or, as Beckermann (2008, p. 87) argued, it is not possible to define terms with a cluster structure explicitly through necessary and sufficient conditions.

In the light of the above discussion, it is obvious that a cluster structure is likely to occur for many (or all?) terms of daily and scientific language, especially in largely descriptive sciences like psychology and psychiatry. The overlap between the minor terms in many diagnoses, so that different diagnoses share the same symptoms, leads to doubt as to whether some diagnoses are actually completely distinct and independent entities, and whether they outline clearly delimitable mental states or not. Examples are attention deficit hyperactivity disorder (ADHD), bipolar disorder, schizoaffective disorder, and schizophrenia. They all bear a cluster structure. They consist of minor terms or symptoms that overlap. This means that many symptoms of one disease are also symptoms which occur in the other disease. Figuratively speaking, if each disease is a link in a chain, then all the links are connected with each other. In this sense, even minor criteria, symptoms, or experiences in near-death states reveal a cluster structure, as the term out-of-body experiences shows.

Incidentally, in the case of near-death experiences, it is better to avoid terms like "symptoms" or "hallucinations", even if the latter only refers to perception without external stimulation, because they have already been burdened with abnormal or psychopathological connotations. But those who survive a clinical death are no less psychically healthy than anyone else. So it seems more neutral to use only the word "experience".

Returning to out-of-body experiences and the cluster structure of this term, we note that it does not only contain a dissociative perception of leaving one's own body; it could also be combined with autoscopy, i.e., "seeing" one's own image or mirror image (see p. 87). The latter phenomenon is itself extremely varied, with autoscopy in the form of a mirror image or suspension over one's own body.

In contrast with the medical diagnoses discussed above, which are often modified as our knowledge increases and, as a result, regularly change classification systems, near-death experiences not only have a problem of cluster structure, but also the problem that several experiences occur *less frequently*. Furthermore, these experiences are *not constant*, in the sense that every near-death experiencer has a different pattern of experiences. And finally, *specificity is not fulfilled*—such experiences are not unique to clinical death, but can also be generated in epileptic seizures, other neurological or psychiatric diseases, drug consumption, and trance, while some even belong to normal psychic life. *For these reasons, the term "near-death experience" does not embrace a specific mental state that is exclusively typical of clinical death.* The term merely connects experiences with a causal event, without the condition that these experiences should *only* occur in clinical death. This is why one should criticise the generally non-critical transition to independence of the term "near-death experience" and its transfer to events that are not connected with a clinical death. The author also disagrees with suggestions (see Greyson 1997) to establish that term in diagnostic classification systems.

In conclusion, it is the author's view that the term "near-death experience" is a categorical mistake. Experiences linked to it are not constant and also occur in various "functional states" of the brain, i.e., clinical death, epilepsy, drug consumption, stress, medical side-effects as we shall see in the following, and many more. In a typical tautological pattern, experiences observed under different circumstances are the proof only of the *term* "near-death experience", but without scrutinising each individual case for the semantic correctness of the term. So the term does not encompass any well-defined entity and should not be used in a strictly scientific sense.

3.4 Exactly When in Clinical Death are Experiences Generated?

Equating "near-death experiences" with death itself is problematic. Every person who has been resuscitated and reported afterwards about their experiences was merely *clinically* dead. Such a situation is characterised by cardiac and circulatory arrest, and thus also an arrest of cerebral circulation, which means that blood flow ceases in the brain. The brain is able to resist such an arrest for a certain time, usually eight to ten minutes (Forster and Ropohl 1989). Under certain circumstances, such as hypothermia, reanimation could still be successful after 20 min.

During clinical death, the brain is not dead but in a state of severe dysfunction! The longer such a state endures, that is, the later a successful reanimation is administered, the greater is the danger of persisting secondary brain damage. For this reason, the argument that near-death experiences are a proof that consciousness could exist apart from or out of the body because in clinical death the brain itself is not functioning just does not work. Despite this, out-of-body experiences or the perception of light are often regarded as such a proof.

Furthermore, there is a problem of chronology. When are near-death experiences generated? At the acme of dysfunction? But why not afterwards in the phase of convalescence? In this sense, reports about null activity in electroencephalography (EEG) during clinical death are interesting. But is non-measurable activity equivalent to a complete loss of function? People who survive a clinical death exemplify the fact that—fortunately—their brain was still alive during clinical death and so could not have been completely functionless. Clinical death is not equivalent to brain death! And even the latter is not defined by loss of EEG activity alone. Various neuropsychological investigations need to be carried out by different physicians at several points of time, demonstrating the arrest of vital functions of brainstem and cerebral blood circulation, before someone can be declared brain dead.

Besides, in EEG, there is a problem of measurement accuracy. How many electrodes should be used? In so-called neuromonitoring in intensive care units, it is often the case that only two frontal electrodes are used. And how long should EEG be recorded? During the transition from coma to agony, slow waves occur more frequently. In the phase of agony, the EEG output could appear as a flat line, but with bursts of high amplitude waves. If the recording time is too short, such wavy patterns will remain undetected and the result could be mistaken for a complete flatline EEG. For this reason, the criteria for brain death comprise among others a standardized EEG recording which must be longer than 30 min!

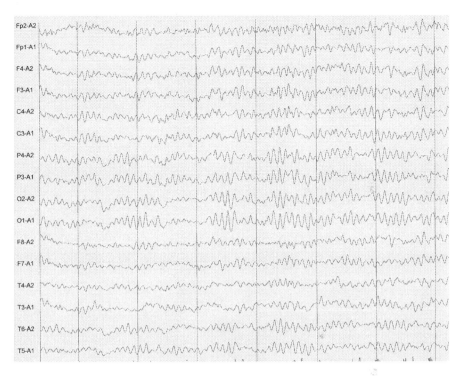

Fig. 3.4 Detail of the normal EEG of an adult

Both EEG's (Fig. 3.4 and 3.5) are only examples to illustrate the type of patterns described in the text. They do not derive from patients with near-death experiences. The EEG itself is not very specific to particular affections of the brain.

An interesting finding was reported by Chawla et al. (2009). They measured high frequency EEG waves in dying patients at the very point of time when the blood pressure was no longer measurable. The authors speculated—as they said themselves—about whether those waves could be a correlate of near-death experiences. However, this idea will remain speculative because their study comprised only seven cases—patients who were so severely ill that life-sustaining intensive care was switched off. In the end, the patients died, so nobody could have figured out which final experiences they had, or whether they even had such experiences. To draw such a conclusion as these authors runs the risk of an ontological proof or a circular argument: the term "near-death experiences" defines itself as the appearance of experiences in the phase of highest agony—in the phase of imminent death. It is therefore understandable to associate a technical finding observed shortly before death—in this case an EEG abnormality—with the appearance of near-death experiences. The final hyperactivity of the brain shortly before death really seems to fit the

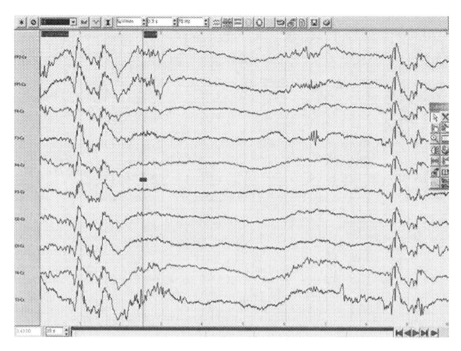

Fig. 3.5   EEG with burst-suppression pattern

theoretical structure that near-death experiences occur in near death, which means in the phase of agony!

But what do we discover here? What did we prove with that conclusion? How do we in fact know that near-death experiences really happen in agony or shortly before real death, at the cutting edge when a life-threatening situation is reversible without secondary brain damage, just in the nick of time? The answer is that we just do not know that! Whenever people who have survived a clinical death talk about their experiences, a certain amount of time has already gone by. It is easy to understand that a physician can talk to a patient only a certain time after reanimation has taken place, after the patient has regained consciousness at the intensive care unit, or been weaned off breathing apparatus, and so on. But in most reports about near-death experiences in the literature, the span of time between the causal event and the report or interview is many months or even years! If we assume the shortest possible span of time, immediately after reanimation, when the patient is awake and able to speak clearly about his experiences, how could we know whether his experiences derive from the phase of agony rather than from the phase of convalescence. Why should they not occur during the slow recovery of brain functions?

At this point we again see how important it is to bring things into question and to investigate how terms or assumptions may simply deliver what they promise! Once more we should have doubts about whether the term "near-death experiences" is scientifically substantial.

3.5 Reprocessing of Experiences Under a Person's Cultural and Religious Views and Publication Bias

An important point to focus on in the following is secondary reprocessing of experiences. Our memory undergoes perpetual change. The content of the memory is constantly matched with emotions, events formerly experienced, and current events which affect us every second; moreover they are connected with wishes, fantasies, and unconscious thoughts. And all this is influenced by culture and world view. Furthermore, nothing is as elusive as memory. This has been demonstrated by studies investigating suggestion in the psychotherapeutic setting. Patients suffering from post-traumatic stress disorder or personality disorders could be made to have false recollections of abuse in early childhood, only through the suggestive intentions of their therapists (Loftus 1997). Similar findings concern witness statements in court. Statements change the more a person is questioned and the further back the event. Often new statements even completely contradict previous ones!

Apparently, latency and the frequent recall and reconstruction of an event in the brain are factors which influence and change details of memory in the course of time. Figuratively, we can imagine that, in every memory recall, the neuronal network is reconnected. The more often such recall occurs, the more likely it is that wrong connections will be made. This leads to incorrect associations of which the person is completely unaware. This modification of memory occurs not only in waking consciousness, but also in dreams. And this also happens in the case of near-death experiences.

At this point we have to face the question of how flowery reports of near-death experiences could lead back to basic neuro-psychological phenomena which subsequently undergo reprocessing. The following case history reported by Lindley et al. (1981) of a man who was resuscitated after a heart attack illustrates the problem. The man reported his experiences afterwards as follows:

…there was a cloud and a male, related to Jesus… the torso was a horse, everything above the torso was a man with wings; sort of like a Pegasus except instead of a horse's head it was a man… and he was beckoning to me… (p. 116)

So here Jesus resembles a mixture of Pegasus and a centaur—a truly uncommon image! Proponents of a supernatural interpretation of near-death experiences often take reports as absolute truths, in every detail! Here, it would imply that Greek myths are right and that they are closer to eternal truths than our beliefs of today! Although we are already far away from science, such scrutinization helps to detect premature convictions of the kind recounted in much of the near-death literature. In this context, it is instructive to note that there is no contemporary image of Jesus at the time he lived. Long after his martyrdom, Jesus was seen as a "good shepherd" with a beardless Apollo-like face. Later he assumed the image of a teacher and philosopher. Because of the latter, he was then portrayed with a beard, standard outward sign of a philosopher. We thus see how our own attitude shapes images. Furthermore, it would have been interesting to know what the man would have reported six months later, to see how far the story would have changed. And it would be interesting to know what he would have told us if he had had no prior knowledge of Greek mythology.

Suggestion also belongs to this part of the discussion. How should I question people when I want to get information about their images of the afterworld or about near-death experiences? Over the years, many questionnaires have been published which should facilitate acquisition and classification of phenomena. Some years ago, Irwin (1987, p. 3), for instance, investigated images of the afterworld in students who had never suffered a clinical death. The students received special questionnaires with questions like these:

A) Heaven appears…
  1. to be a city surrounded by high walls and entered via large golden or pearly gates,
  2. very much like Earth, with streets and buildings, villages and towns,
  […]
  6. to be a place bathed in perpetual sunshine.

Other parts of the questionnaire investigated whom one could expect to meet in heaven and whether one would want to move there. Students were asked to imagine being dead and that they were in heaven. After they had imagined this, they had to compare their "experiences" in several items on the questionnaire. On the one hand, the investigation revealed the interesting finding that stereotypical images of the afterworld do not exist and that sociocultural influences are important; on the other hand, suggestive questions have a striking effect which weakens the meaningfulness of the study. With regard to near-death experiences, assessment scales like the 16 point scale of Greyson (1983) which has been used for studies and investigations in many publications and which asks for cognitive, affective, paranormal, and transcendent aspects, also carries with it the problem of suggestive questions.

Another problem is to circumscribe "real" near-death experiences, i.e., in clinical death, rather than similar experiences in other situations or diseases. Hoepner et al. (2013), for instance, pointed out that, in their study, people with ictal autoscopy (i.e., autoscopy which occurs as a symptom of an epileptic seizure) also got high scores on the near-death experience scale.

Similar problems accompany a recent study which pointed out special "characteristics of near-death experience memories as compared to real and imagined event memories" (Thonnard et al. 2013). The authors investigated 21 patients who "suffered from an acute brain insult and recovered from a coma" (p. 5). Using Greyson's near-death experience scale, they divided their sample into three subgroups and a control group of healthy subjects. The three subgroups of patients were as follows: those who reported memories of a near-death experience, those who reported memories associated with their coma and intensive care period, but without near-death experiences, and those without any memories of their coma. After assessment of these groups by the Memory Characteristics Questionnaire (MCQ), the authors described *"a significantly higher amount of memory characteristics for sensory, clarity, self-referential information (SRI) and emotion categories for NDEs* (i.e., near-death experiences, a.a.)…" (p. 3). They concluded that *"what makes the NDEs 'unique' is not being 'near death' but rather the perception of the experience itself."* (p. 4)

So far, so good. But the question arises as to why the term "near-death experiences" was used when it was obvious that none of the patients had had a clinical death? They "only" suffered from a stroke and ensuing coma! Worse, if we look more closely at how the subgroups were created, yet more doubts arise. Among others, Greyson's near-death experience scale (IANDS 2011) comprises items like:

Did you have a feeling of peace or pleasantness?
    0 = No
    1 = Relief or calmness
    2 = Incredible peace or pleasantness
Did you have a feeling of joy?
    0 = No
    1 = Happiness
    2 = Incredible joy
Were your senses more vivid than usual?
    0 = No
    1 = More vivid than usual

2 = Incredibly more vivid
Did you seem to enter some other, unearthly world?
    0 = No
    1 = Some unfamiliar and strange place
    2 = A clearly mystical or unearthly realm
    [...]"

The modified version of the Memory Characteristics Questionnaire which was used by the authors (Thonnard 2013, p. 4) inquires once again about the senses, clarity, and self-referential information (to remember what I did during the event). In conclusion, there is a high degree of overlap between the items in the two tests. As a consequence, patients who had been defined as having had a near-death experience by the Greyson scale would necessarily have higher scores in the memory questionnaire! So the authors used erroneous circular reasoning. They found certain characteristics in a subgroup, but they established the subgroup using those very same characteristics! What does the study tell us about near-death experiences? Nothing! Again we see that experiences with clarity and an emotional content occur under different circumstances, such as after a stroke and coma. But those experiences are not necessarily connected with a certain kind of event (disease), and nor do they have to occur in all subjects. The study reveals how an initial mistake beginning with an uncritical categorisation or definition of the term "near-death experience" leads to bias! What research could learn from this is that it is worth scrutinizing such popular terms as "near-death experiences" before starting a study.

Besides, the Greyson scale defines near-death experiences only at a phenomenological level. The question is: what do we actually measure? A more or less arbitrarily established arrangement of symptoms that we use to define the term "near-death experience"? What knowledge does the term provide us with concerning the multiple evocation of mental states? If "everything" is a "near-death phenomenon", what new knowledge about its causes do we acquire? In other words, there is no clarity about any core symptoms which the 'diagnosis' near-death experience would require, and nor do these experiences comprise unique characteristics that would distinguish them from other experiences which occur in other circumstances than clinical death alone. As for questionnaires and scales, the cart was put before the horse! First we define what we want to investigate and then other theoretical constructions can build up on this for further research. If the first part fails, as has already happened with the nebulous term "near-death experience", research cannot deal with such broad concepts, but can only focus on immediate components.

What makes research difficult is that, in clinical death, malfunction alters a "mature" brain with individual thoughts and feelings embedded in a certain cultural and religious background. Thus the configuration of near-death experiences gives a hint of the cultural and religious world view of the person affected. In short, we would need a Kaspar Hauser to distinguish interpretation and secondary reprocessing from whatever were the neuro-psychological core phenomena. Kaspar Hauser had a most unusual fate. One might say that he was a feral child. Hauser (born around 1812, died 1833, Ansbach) was found one day in Nuremberg at the age of 16. He was seriously retarded and had only a rudimentary ability to speak. It is supposed that he had lived somewhere in isolation for many years. Later on his educational attainment improved quickly. There has long been debate about his origins, including rumours that he was a prince who was exchanged after birth with a sick baby, the intention being that the latter would soon die, leading to a change in the succession of the Baden dynasty.

In German behavioural biology, the term "Kaspar-Hauser experiments" stands for investigations of animals that have grown up completely isolated from conspecifics, the aim being to investigate whether a certain kind of behaviour is determined genetically. Could such a Kaspar Hauser be a model, or at least a *gedankenexperiment*, for our discussions about near-death experiments? If such a patient experienced a clinical death, would he tell us only about the neuro-psychological basis phenomena, given that he has neither a cultural nor a religious world view? Would he tell us about such pure experiences when all possibility of secondary associations, reprocessing, and suggestion has been eliminated? But then how could he express himself? Language is a medium of communication, and language is always exposed to cultural influences. Behind every word, there is already a thought and much more. There is already a concept of the term which is expressed. Thus a *communicating* Kaspar Hauser would no longer be a "real" Kaspar Hauser.

The influence of worldview is shown in an interesting study involving both the Eastern and Western federal states of Germany (Schmied-Knittel 2006). Two decades ago, Germany consisted of two different states, subject to different cultural influences induced by the disparate political systems on either side of the iron curtain. In short, in the East, the atheist worldview became a doctrine, whereas in the West, religious concepts prevailed. Even two decades after reunification, differences in attitude to religion and spirituality are obvious, and this affects near-death experiences too. In the study, Germans from both East and West were interviewed with regard to near-death experiences. The overall frequency of near-death experiences was four percent. The author concluded that there was no connection between clinical death and the occurrence of near-death experiences. In the sample, *"surely existential, but by no means life-threatening circumstances"* were designated as causes of near-death experiences. Most of these circumstances involved *"road and workplace accidents, operations, heart attacks, and other acute diseases"* (Schmied-Knittel 2006, p. 241).

What is reported about the individual experiences? Out-of-body experiences, light phenomena, and feelings of having been in another world were reported more frequently in the West. The Easterners reported more often about tunnel phenomena. In the West, positive experiences were more frequent than in the East, where negative ones dominated. Concerning not the East-West separation, but rather a gender contrast, men more frequently reported negative experiences, while women were prone to positive ones. On the whole, people from the West mostly interpreted their experiences either in a traditional religious way or through parapsychological or esoteric concepts, or concepts from Eastern religions, whereas in the East of Germany, non-religious explanations prevailed. In Germany, more and more people are open to esoteric concepts. Until today, the East and West have differed in detail. The western population is much more prone to such concepts. For this reason, the above-mentioned investigations with regard to near-death experiences should come as no surprise. One would expect that, if such interviews were carried out again in a couple of years, the difference between the East and West would have shrunk or even vanished. Presumably then, spiritual explanations of near-death experiences would become more and more the focus of attention. Since the German reunification in 1990, churches have been losing members at a dramatic rate. In contrast, spiritual and esoteric movements are increasingly dominating the daily lives of people in both former parts of Germany.

Another discussion point is the number of publications about near-death experiences, and also their intentions. In recent years, the topic has been widely debated in the "Western" world, both scientifically and in the media. It may be that the high number of near-death experiencers compared with the relatively small number of reanimated people reflects the trend in favour of esoteric movements in the West—having a near-death experience, even independently of enduring clinical death, enlarges one's horizons and gives insights into eternal truths! So the frequency of near-death experiences, or certain sorts of them, is certainly biased by sociocultural changes! In addition, this impedes attempts to distinguish between different types of experiences because of the enormous distortion factor which the religious and cultural background introduces. This further hinders the search for a possible neuropsychological core.

The need for spirituality is unabated in Western society. Near-death experiences play a prominent role because they appear to lie at an intersection point of science and spirituality, tempting those who would seek a proof of the existence of God or the supernatural. How nice it would be to get a reliable proof that we would somehow continue to exist after physical death—a proof that is more than just a belief, which is all that a religion can provide for its devotees. This would imply that we could worry less about threats to our existence. It would represent a tower of strength amidst the uncertainty of life! The anxieties of the present age resemble those of Hieronymus Bosch's day, with its apocalyptic visions: climate change vs. the Little Ice Age, terrorism vs. the threat to Europe by Turkish armies, and avian flue vs. the plague. Small wonder that near-death experiences have been so widely discussed over the decades since Moody's famous book and—as we have shown—even long before.

3.6 Religious Contradictions from a Philosophical Point of View

Not all people who survive clinical death report near-death experiences. Quite the contrary, only about one fifth in prospective studies (Mobbs and Watt 2011, p. 448) to one quarter (Schwaninger et al. 2002; see Martens 1994) report about such experiences afterwards. In other words, 75 to 80 %, the vast majority of clinical death survivors, do not report near-death experiences. In retrospective studies, the relation is approximately fifty-fifty (Mobbs and Watt 2011, p. 448). That difference between prospective and retrospective studies is already rather surprising. One explanation could be that prospective studies do not allow the patient so much opportunity for secondary interpretation and reprocessing of the event. In the most favourable circumstances, the researcher performs the interview immediately after a patient has recovered from reanimation. In contrast, retrospective studies might typically be performed as follows. A researcher checks charts of former patients to see whether they fulfil certain criteria—in our example a criterion would be survival of reanimation. Then the researcher gets into contact with them to arrange an interview. The difference is that here the risk of distorted utterances is greater because of the time lapse between the clinical event and the date of interview.

But whatever the proportion of "non-experiencers" is, this fact alone should raise a religious and philosophical question in those researchers who regard near-death experiences as a proof for the existence of the supernatural: What about the other three quarters of survivors? Why is it that they get no insight into the afterworld? And why do such experiences arise more often in circumstances that have nothing to do with the dividing line between life and death?

Discussion about near-death experiences revives an old philosophical dispute—the mind-body problem. Do mental states exist completely independently of the brain or are they always connected with brain function? Out-of-body experiences in particular are taken as a proof for the independent existence of mind and body by those who interpret near-death experiences in a spiritual way. A mind which exists completely independently of the body is, in philosophical terms, equal to a form of parallelism. But is parallelism compatible with the findings of neurosciences?

To answer that question we have to take a step back and investigate what the word "mental" itself could mean. But that question is tricky. According to Beckermann (2008), mental states consist of sensations and so-called intentional states. The latter are attitudes such as convictions and attitudes with a conative component, i.e., containing an impulse to act, and affective components. Taken together, such attitudes determine what we call personality. Brain diseases like frontotemporal dementias in which frontal and temporal lobes are affected, in contrast to Alzheimer's disease, lead to massive alterations of personality. Personal convictions, feelings, mood, activation, structured planning and action, organisation of activities—that is, affective and conative attitudes—are drastically changed, with the result that personality is destroyed. Even in maniac states or severe depressions, activation and structured action are also impaired.

It is clear that mental states require "substance" (i.e., "body", brain), even though that does not necessarily mean a pointwise dependency. We thus have a contradiction with parallelism.

A loophole here might be the occasional, irregular, and unexpected intervention of the supernatural in our everyday lives. In this sense, the widely used statement that God moves in mysterious ways is a cheap way to avoid critical argumentation. But at this point, we definitely leave the narrow road of science and move into the unconstrained sphere of belief. This is precisely why there can be no scientific proof of the supernatural, but also no counterproof. Science and belief are distinct things.

From a religious point of view, it is also worth asking whether everybody could potentially get insight into the afterworld. If so, how important are praying and belief, and how important would one's own way of life then be, in the sense of moving a person nearer to God? Is intensive contact with the supernatural a blessing that anybody could achieve during his life? Or should such wishes be regarded as an aggrandisement that would be undignified for man, because of all his flaws and shortcomings? Should we be more modest, trying only to make the world better, rather than claiming insights that we would never be able to understand? Or does the transcendence of a prayer or a ritual differ gradually but not qualitatively from a spiritual "near-death" experience?

Reports about near-death experiences mostly derive from the Western world and thus have a connection with Christianity. No wonder then that many people consider their experiences, especially visions of light, as a minis-

LESSING.

Fig. 3.6 Portrait of Gotthold Ephraim Lessing, copper engraving around 1820 by Ferdinand Bollinger

try of Jesus. But if a scientist takes such interpretations for reality, this means abandoning the distance and objectivity that are seminal for scientific work. It also amounts to claiming to have found the real ring in the parable of the ring told by "Nathan the Wise".

This play by Gotthold Ephraim Lessing (born 1729, Kamenz; died 1781, Brunswick) (Fig. 3.6) was first published in 1779 in Berlin, with the first English edition in 1805. The plot unfolds in Jerusalem during the third crusade (1189–1192). The Sultan Saladin asks the Jew Nathan, who is famous for his wisdom, which is the right religion. Nathan answers with a story, the parable of the ring, which was not Lessing's own invention, but reflects much older literature. It is about a ring which has always been bequeathed from the father to his most beloved son over generations. One day it happens that a father has three sons but does not want to favour one of them over the others. For this reason, he orders two exact copies of the ring. Later the sons appear before a court, but even the judge cannot define which ring was the real one.

Nathan said: "What I mean
Is merely an excuse, if I decline
Precisely to distinguish those three rings
Which with intent the father ordered made
That sharpest eyes might not distinguish them."
(Lessing 1991; Scene7, p. 233)

Saladin subsequently befriends Nathan. The rings symbolize the three mono-theist religions Judaism, Christianity, and Islam. The whole play is an appeal to religious tolerance.

On the other hand, it is also understandable that every religious person could only consider their own religion to be true. This is a precondition for religious life. But even here, we cross the borderline between science and religion. In reports of near-death experiences, it is necessary to take into account the fact that everyone has been influenced to some extent by religious concepts. Even an atheist knows, for instance, about Jesus, and has some kind of image of him from pictures or visits to churches. So Jesus has become part of his mindscape, if only unintentionally.

As has already been described (p. 56), Hindus often experience the god Yama in near-death. Others studying the experiences of children report that they "saw" a teacher, a physician, a conjurer (Kasten and Geier 2009), or a "golden haired angel" (Morse and Perry 1992, p. 23), horses (ibid, p. 52), scenes from school (ibid, p. 71), children playing (Lindley et al. 1981, p. 112), and a deer (ibid). And so we come full circle. These explanations touch upon both our previous thoughts about what we referred to as the "Kaspar-Hauser model", and secondary reprocessing of experiences in the context of a person's cultural and religious views, which constitutes unconscious (and conscious) interpretation of basic neuropsychological phenomena.

Similarly we have to ask how the increase in reports on TV and radio and in books shapes certain views of those who may subsequently interpret such a near-death situation in the case of a medical emergency. Certainly, this development is responsible for the fact that the term "near-death experience" is much more widely used than what it actually stands for—its occurrence in clinical death. On the contrary, the term is also used for experiences which have nothing to do with clinical death or any danger at all. Here, we must also criticise the psychotherapeutic specialisation to "near-death experiences" which attempts to use them as a form of diagnosis, because this leads to an overinterpretation of those abnormal experiences that are not connected with clinical death. For instance, people who had formerly been diagnosed with dissociative disorder or drug consumption, or even psychically healthy people, become rededicated! But such rededication does not provide us with any deeper insights. It is merely a renaming in accordance with the latest fashion. Furthermore, such developments go against scientifically substantiated "diagnostics" which critically review social and media tendencies. Similar problems are well known with diagnoses such as mobbing, stalking, animal-hoarding syndrome, and so on, without asking whether behavioural problems occur only in a phenomenal sense or whether they involve different and definable mental entities. Nevertheless, such diagnoses are becoming more and more frequent in psychological and psychiatric terminology.

3.7 Near-Death Experiences in Esoteric Movements—A Modern Gnosis?

Esoteric movements go beyond a religious belief in a god. The word "gnosis" comes from the Greek language and stands for knowledge. It is more than a belief in supernatural forces—it provides a person with insights into the proceedings of the afterworld. These insights could be attained with the help of pseudo-scientific concepts. Approximate knowledge of scientific procedures or scientific theories, especially those that are not yet completely understood, is somehow combined to build up a system in which the existence of the supernatural can be demonstrated by scientific means.

An example of a scientific procedure used to prove the existence of the supernatural in near-death experiences is the EEG paradox. There are some reports of near-death experiencers who have had an electroencephalography (EEG) with null activity. As discussed on page 62, the inadequacy of the measurement itself—only a few electrodes, short time of measurement, recording of only superficial brain activity—are edited out. Instead, null activity is taken as a proof that the mind can exist independently of the brain. For how should near-death experiences be generated by the brain if it is not working at all? And neither do these accounts address the issue of how we know that near-death experiences occur in highest agony rather than in convalescence.

Another theory often used for esoteric explanations is quantum theory. Even physical effects of light or the speed of light merge in esoteric literature as explanations for the soul. Such constructions promise knowledge about "consciousness beyond life", "how consciousness survives death", or a "proof of heaven"—to quote some headings of contemporary and influential near-death literature.

Other features are interesting:

* Experiences are regarded as a proof of the existence of the supernatural in exactly those forms in which they appear. Each detail of the experiences is

Fig. 3.7 Portrait of Bernard Le Bovier de Fontenelle, copper engraving around 1820 by Wilhelm Devrient (born 1799)

regarded as authentic. Thus, for many, their experiences are also a proof of the correctness of their religion. Another feature is a corporal view of the mental. Even if souls seem to remain after death, in near-death experiences, decedents often appear as people, with bodies. The typical composition of spiritual knowledge by means of notions pertaining to our human and material world had already brought critics to the scene hundreds of years ago. French philosopher Bernard Le Bovier de Fontenelle (born 1657, Rouen; died 1757, Paris) postulated in his 1684 script "De l'origine des fables" [The origin of myths] that the ideas inherent in myths are recreated on the basis of already well known entities (Fontenelle 1989) (Fig. 3.7). Immanuel Kant thus argued against the possibility of proving the existence of God or the soul "...*from a theoretical point of view*":

And there is a perfectly intelligible reason for this, since we have no available material for defining the idea of the supersensible, seeing that we should have to draw that material from things in the world of sense, and then its character would make it utterly inappropriate to the supersensible. (Kant 1953, p. 139)

* Often today esoteric views resemble those spiritual concepts previously borne by dreams or apparent death. One of these could be called world spirit. In esoteric concepts, it is often combined with "carrier substances" such as light or merely energy. When liberated, the soul (spirit, mind) plunges into the world spirit and thus obtains insights into the afterworld and life both before and after our time. Reports on divination during near-death experiences are not uncommon. The soul is released at the cutting edge between life and death, so clinical death carries great importance for such concepts. But on the esoteric scene, the latter plays an ever more minor role. In the esoteric movement, every insider is able to get spiritual insights by various means and under a multitude of conditions. This is why in such concepts it is commonplace to transfer the term "near-death experiences" to circumstances well removed from any life-threatening situation. It is becoming more and more fashionable to have one's own near-death experience—as a spiritual illumination. The concept of world spirit is ancient, having its roots in Pythagorean philosophy and Eastern ideas of reincarnation, and finally becoming incorporated into the gnostic dogmas which competed with early Christianity 2,000 years ago. Here, god stands for world spirit, and during creation he gives spirit gradually to the animate and the inanimate. This gradual distribution of spirit is called emanation. The aim of every individual should be a reunification of the divine part of his soul with the big picture. Mani (see above) also argued along these lines. Many theologians classify his dogma as gnosis. It seems that many points of discussion in near-death experiences merely reflect much older problems.

Why have near-death experiences always offered broad spiritual views?

* Survival of clinical death is an incisive event in life which raises essential questions of human existence. And this includes spiritual views, too. Science cannot and should not attempt to solve these questions. But science can certainly help to scrutinise religious dogma.
* In contrast to dreams or other phenomena of daily life, near-death experiences can be more easily connected with the afterworld or the supernatural because they mark a boundary of worldly existence which is accepted by both the general public and scientists.
* Near-death experiences often reveal personal wishes (Fig. 3.8). These include proof of one's preferred religion, meeting with beloved decedents, or redemption of worldly burdens in an afterworld full of delights yet to

Fig. 3.8 Tomb at the Trinitatis cemetery in Dresden with the German inscription "Wiedersehn," which means "see you again" and illustrates the wish to meet the dead person again

come. In short, near-death experiences provide insights into cosmogony— the origin of everything—and also eschatology—the fate of everything. So generally speaking, apart from a few "negative" near-death experiences, they diminish the fear of death. It is well reported that near-death experiences give new insights to the affected people. Against a background of rapid popularisation of the topic in the media, this might explain why many people, apart from those who have survived a clinical death, seek their own personal near-death experience. Such a *proof* of the afterworld—could one's own experience be wrong?—is more valuable than a mere belief of the kind that a moderate religious attitude would provide.

* Arguments often resemble a kind of "god-in-the-gap" reasoning. The gaps which science is unable to fill out with answers are regarded as a proof of the existence of the supernatural. This also happens in near-death experiences when the discussion seeks unique characteristics or experiences that are not explainable. Neuropsychological models are often taken to fail because they cannot provide a person with a point-for-point representation of mental functions. Furthermore, all mental states differ at least in the

details. But why should God only appear in the gaps and not in the perceptible world? Esotericists often want to challenge the sciences to explain everything, every detail. Safe in the knowledge that this will never happen, they take this failing as a proof of their esoteric concepts—without scrutinising them to see whether there might be a better, rationally explainable alternative. But what would such a demand involve? To resolve the conflicts between knowledge and belief, man must explain everything, even the essential questions of destiny and the meaning of life. And what would that involve? Provided with such a complete knowledge, man would lift himself to the same level as God! Being godlike, he then could analyze or even refute God? But how would that work? It is simply a contradiction in itself and what is more, it is boastful. Man has to accept the imperfections of life and the incompleteness of knowledge! On the one hand, that makes our life interesting because research provides us with new insights all the time. But on the other hand, it allows different world views.

At present it is difficult to predict the impact the esoteric approach will have on our society. Will purported insight into "last things" of the kind claimed by spiritual near-death experiencers lead to new concepts and a new movement that will not only compete with, but completely change "traditional" religions? As we see in history, society shapes religion and vice versa.

4

Medical Theories

4.1 Do Neuropsychological Basis Phenomena Exist?

From the discussion so far, two main problems crystallise out. These two problems will be important for establishing a medical theory.

* Firstly, are reports from near-death experiencers integral, "primitive" experiences which date from the most intimate temporal connection to the phase of clinical death? Or are they merely a secondary interpretation of some neuropsychological basis phenomena?
* Secondly, do these experiences fall into a constant pattern which could be detected in everyone who had near-death experiences?

The first question has already been answered sufficiently in the above discussion. The influence of a person's world view or *weltanschauung* on the embodiment of reports is a fundamental fact. But there are nevertheless hints that core experiences or neuropsychological basis phenomena might in fact exist. In support of this claim, there are reports from children who have experienced tunnel phenomena, whereas children have not yet created any ideological belief. Morse et al. (1986) interviewed children from the ages of three to twelve who had survived cardiac arrest or deep comas. Seven of the eleven children reported tunnel phenomena and autoscopy. But even in this study, we come up against the fact that there can be multiple realizations of the same mental states: the authors investigated both cardiac arrest and coma, two different triggers that resulted in the same experiences! And as usual in near-death studies, the number of patients is rather small.

But to answer the question of basis phenomena, we also have to take into consideration reports about visions of lights or structured visual phenomena such as "spiderwebs" or zigzag patterns. We could indeed assess the latter as neuropsychological basis phenomena because they are concomitant with brain malfunction. Visual phenomena include unstructured lights called phosphenes, whereas structured visions with similar geometric figures are called photopsias. The latter is reported in two interesting case histories:

In the first case, an Uzbek woman we interviewed (Engmann and Turaeva 2013) reported feeling that her body was moving, together with noise and a zigzag pattern. She had had a hip operation and was reanimated with electro-shock therapy. She remembers that, before narcosis, she had been laid on the operation table and received an injection. After that, it was like a dream. She saw lights as in the operating room, felt that she was up, and saw the doctors doing something. Afterwards she had a feeling of moving around something like a table. She heard like tick-tick-tick, then saw a zigzag pattern which she depicted as a row of tents like this: ΛΛΛΛΛΛ (p. 3)

Here we already have to face an uncertainty. Which of the reported experiences really derive from reanimation? The feeling of something being injected, of having a dream, seeing light, and even the out-of-body experience could still be connected with the *beginning* of anaesthesia, when the drug is just starting to take effect! The light could be the light of operating room and the out-of-body experience could be caused by the anaesthetic.

In the second case (Lindley et al. 1981, p. 111), a man was resuscitated after a heart attack. He afterwards reported:

> The more I concentrated on this source of light the more I realized that it was a light of a very, very peculiar nature... It was more than light. It was a grid of power... If you could imagine the finest kind of gossamer spider web that was somehow all pervading, that went everywhere.

Amazingly such structured visions also occur in other circumstances.

A man was admitted to hospital with a left-sided palsy of the arm and leg, but which was only of short duration. An MRI scan revealed no stroke. The good prognosis of the palsy and the MRI scan support the diagnosis of a reversible ischemic neurologic deficit (RIND). The man explained that he saw people on the wall who talked to him, and magnificent carriages with horses, also on the wall, followed by horrible spiders and spiderwebs together with flashes of light. These latter visual phenomena are a hint of some functional involvement of the occipital lobe structures. The man was objective about his misperceptions and clear of mind. In the course of time, the intensity of the hallucinations weakened. (Besides, palsy and visual phenomena are a sign of multiple affections of different brain regions which often occurs in embolic strokes. I don't know the subsequent course of his hallucinations. Had they persisted, they could also be regarded as a stroke even though not visible in the MRI scan).

At any rate, here we have a report of scenic but bizarre visions which also occur as near-death experiences. An example of such bizarre experiences can be found in Lindley's publication (Lindley et al. 1981, p. 116) about the man who was accidentally electrocuted and later claimed to have seen Jesus in the form of Pegasus (see the discussion on p. 65).

The difference between the man who saw carriages and the one who imagined Pegasus is that, in the first, the visions were visual misperceptions that continued when he was awake, while in the latter, we have a recollection of a previous event that was closely connected with the accident. But a similarity is the graphic and somewhat quaint content of the visions.

Once again, we observe:

* the fact that similar experiences could be triggered by completely different events than clinical death,
* the wide variety of "near-death experiences",
* the influence of personal recollections and, of course, interests.

Graphic experiences like the initially mentioned film-clip review of one's own life also occur, but such phenomena are rare, and less frequent than the perception of light, for instance. Furthermore, the idea that it is the entire life which unspools like a film in a projector has never been verified. Usually, we find a selection of scenes with autobiographical content. These complex and structured visual hallucinations are in fact typical when problems occur with the integration of the functions of the occipital, parietal, and temporal lobe structures.

Tunnel phenomena could be explained by a central narrowing of the visual field or over-representation of the fovea in that visual field in the occipital regions of the brain. The *fovea centralis* is an area of the retina which is located in the macula. It is the point of sharpest vision. This structure has a counterpart in the occipital brain, whence it has its own representation there. But while the fovea centralis in the retina of the eye constitutes only a comparatively small area, its counterpart in the brain dominates the primary optic area (the visual cortex) to the extent of occupying four fifths of it (Trepel 1999). If there is a severe impairment of the brain, as happens in clinical death with the lack of nutrients and oxygen caused by arrest of the blood circulation, it is quite conceivable that cell malfunctions will occur. Then the biggest area will make the biggest impact. The fovea cell junctions in the primary optic area occupy the biggest area, so they will control this kind of symptoms or experience. This might be an explanation for the light tunnel.

The feeling of moving through a tunnel also requires that the person should feel that her or his body is up and moving, so brain regions responsible for body position might be involved too. The pathophysiological correlate of the tunnel view in the context of near-death experiences is a controversial issue in the literature. One problem is that this phenomenon cannot be scientifically

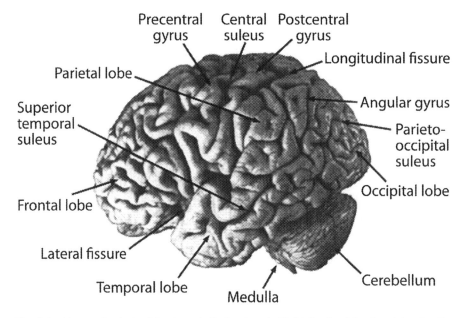

Fig. 4.1   Human brain in side view. *Left*: forehead. *Right*: back of the head (occiput)

investigated directly in clinical death, so we can only compare similar experiences in other mental states, e.g., diseases or G-LOC syndrome, and then conclude to the near-death situation.

Tunnel phenomena sometimes occur as an aura symptom in migraines (Engmann 2011a). Although migraine is a not a rare disease, it is still not clear what causes it. Several hypotheses are in competition. The one favoured now is that of spreading depression, a wandering depolarisation of the cortex mediated by potassium ions. At the onset, hyperactivity is assumed to be located in the optic area (visual cortex), which explains why visual phenomena, and in particular zigzag patterns, are typical at the inception of a migraine—the so-called aura. But an aura occurs only in 20 % of all migraine attacks. There remain several open questions. It would be a misinterpretation if we now tried to argue that the same process occurs in the tunnel phenomena associated with clinical death, but we can deduce that this phenomenon presumably originates from structures of the occipital cortex. Furthermore, there are reports of the tunnel phenomenon under ether narcosis (Dunbar 1905, p. 75), which was common around the year 1900. Moreover, abuse of ether was frequent. Even here the question rises as to whether the same phenomena indicate the same originating anatomical structures?

After a near-death event, some report that they have "seen" friends or relatives, or others closely related to them, and mostly people who have al-

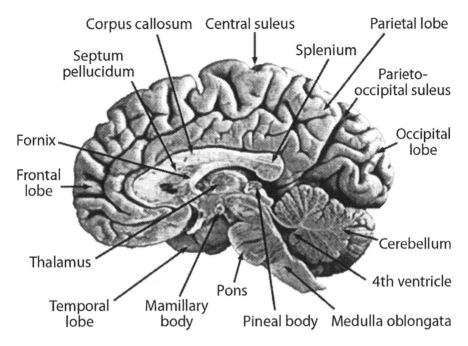

Fig. 4.2 Cut through the brain. The limbic system includes among other things the hippocampus, fornix, cingulated cortex, amygdale, and parts of the thalamus.

ready died. One question is: Did they recall direct visual recollections of those people or did they experience unclear hallucinatory elements which were later subject to interpretation. A hint that we should consider the latter might be the uncertainty in the reports of a kind that is often uttered by the experiencers themselves: it looked like, a sort of, a thing like—compare with reports above! Note that such visual experiences of intimate acquaintances are also well known in the case of occipital lesions in which connections to the temporal and occipital brain are altered (Kim et al. 1993; Kölmel 1993), and have also been generated in cortical stimulation experiments (Penfield and Perot 1963). Subjects "see" not only people or objects, but also film-clip sequences. In such cases, these visions have a scientific name: complex visual hallucinations.

Let us say a few more words about the often evoked dualism of basis phenomena and secondary reprocessing. This brings to mind postulates from the popular book *The interpretation of dreams* by Sigmund Freud (born 1856; Freiberg in Mähren/Příbor, died 1939, London). A "latent" near-death experience—in our view a neuropsychological basis phenomenon—changes by secondary reprocessing under a person's cultural and religious views into a "manifest" near-death experience, which means that it has now become equipped with a flowery description of individual and cultural diversity. Sigmund Freud postulated such a secondary reprocessing for dreams, making it responsible for the fact that some contents of a dream sink into oblivion, while other "material" is reinterpreted and transformed. But how could such psychodynamic aspects of a near-death experience—not the near-death experience itself!—be associated with dreams? This is a reflection I would like to bring up for discussion.

Another important question is how "resistant" these experiences are? If people are once more interviewed about their near-death experience, or twice or three times, to what extent would their experiences differ in the course of time?

A positive perception of near-death experiences may reveal how much wishes play a role in interpretation—unless such positive feelings are generated in the limbic system, a net of different structures in the brain responsible for feelings of joy, warmth, and love, but also grief and anxiety. There are also reports (Irwin and Bramwell 1988) of near-death experiences involving negative feelings of grief and anxiety, but they seem to occur seldom. Are people inclined to give a positive explanation of their experiences—in relation to the situation in which they occurred—because they provide answer to one of the most important questions: What comes after life? Would that require there to be only a positive interpretation? In this sense, the above-mentioned reports of Thai subjects, who seem rather to experience frightening things, could reveal such a cultural and religious preoccupation. Or is it all caused by a somewhat random affection of the limbic structures in clinical death, which provides people in near-death with a certain kind of feelings?

Whatever the case may be, the limbic system is always a mediator of feelings—as it is in "normal" life and during a possible reinterpretation, so it also might be in clinical death. The only question concerns the exact point of time at which it comes into being. The same goes for the occasionally reported feelings of familiarity which accompany people when they "travel" through the tunnel and into the afterworld.

In addition, lifelikeness and sustainability are often taken to be distinguishing features of near-death experiences. But this is not a unique characteristic either; such lifelike recollections are also found in people who have experienced traumatic life events. The acme of such realism occurs in the so-called flashbacks, when complete scenes are re-experienced. On the one hand, clinical death really is a traumatic situation, so recollections of it have a distinct representation in memory as compared with events from daily life. On the other hand, the conclusion that near-death experiences are related to post-traumatic stress disorder in general would be premature—this will be discussed in detail later. But we see once again that the same psychic functions of the brain could be evoked in different circumstances, so this is one more example of what we call multiple realizations of mental states.

Let us continue the discussion of the interplay between "interpretation" and "neuropsychological basis phenomena." What about meeting with decedents? Who really wants to accept that the death of a loved one could be inevitable and without possibility of return? It is noted here that personal meetings with decedents, and especially friends and relatives, are a central theme among the beliefs in the hereafter propagated by most religions. The same goes for out-of-body experiences, which also have their origin in a malfunction of the brain, as we will see in more detail now.

4.2 What is Special About Out-of-Body Experiences?

The frequency of out-of-body experiences also varies significantly from one study to the next, fluctuating between one fifth and two thirds of all near-death experiencers. They are an interesting and much discussed phenomenon. Proponents of supernatural explanations use this suspension of the mind over one's own body as a proof for dualism, taking it to imply that body and mind exist independently from each other.

Out-of-body experiences are related to autoscopy. This is a phenomenon in which people have a visual hallucination of themselves, i.e., they see and recognize their own image in front of them. But many descriptions of near-death experiences—and not only this situation—contain more than autoscopy alone. Autoscopy often occurs in migraine and epilepsy, and usually consists of a visual hallucination of one's own upper body parts, or only the face. Those who experience autoscopy often report that they have the impression that such a mirror image appears at arm's length. In rare cases, the mirror image is multiplied, with the mirror images overlapping. To imagine this, one can go into an elevator in which the cabin walls are made of mirrors. If one looks into one mirror, there will also be an endless sequence of mirror images in a row. Such multiple images are called polyope heautoscopy (Brugger et al. 2006). In the literature, autoscopy is often connected with strokes in the occipital region. For instance, Zamboni et al. (2005) described a 30 year old woman who suffered from bilateral occipital lesions and additional lesions of the basal ganglia. She reported seeing herself as a mirror image. The apparent distance of her hallucination varied every once in a while and the mirror image imitated all her movements. It was transparent, which meant that she was able to look through it and recognise other—really existing—people or things. Bhaskaran et al. (1990) wrote about a 60 year old patient who had an acute ischemic stroke in the right occipital lobe. He hallucinated a person whose face and upper body parts he identified as his own.

It might be that these phenomena do not have their origin in the occipital brain lesion alone; moreover, *integration* of different brain functions seems to be impaired. A further hint here comes from investigations by Thonnard et al. (2008). By medical imaging, they showed that the temporo-parietal cortex is involved in provoking out-of-body experiences. Britton and Bootzin (2004) analysed EEGs of people who reported having had near-death experiences with transcendental content. Compared to a control group, these people more often showed epileptiform potentials in electroencephalography of the left temporal lobe. The finding of left-sided temporal alterations is interesting here, since findings in other studies often refer to the right. Recently, Hoepner et al. (2013) investigated patients who experienced autoscopic phenomena during an epileptic seizure. EEGs revealed *"epileptic dysfunction at the temporo-parietal junction or its neighboring regions"*. (p. 742) MRIs of the five cases revealed abnormalities and lesions in these regions, including among other things local malformations of the cortex and a surgical defect. In this tiny sample of five patients, it was mainly the left hemisphere that was affected, but one subject was also affected on the right side.

One should ask here whether the autoscopy is caused by activation of memory of one's own image, or whether altered recognition of faces might be the cause. The latter has something to do with the right temporal and right parietal parts of the brain (Kolb and Whishaw 1990). That would imply that a hallucination which looks like a person is not necessarily the person having the hallucination, but rather that this person mistakenly interprets it as their own body or face. We might also ask whether psychodynamic aspects play a prominent role. As Kölmel (1993, p. 260) put it:

> The blank visual field, excluded from external reality, serves as a screen on which images from the individual's inner reality are projected. Therefore, the contents of complex visual hallucinations should be interpreted psychodynamically in the same manner as dreams.

Some affected subjects report experiences which resemble a so-called depersonalisation—the feeling of being apart from one's own body, but without autoscopic phenomena. Such depersonalisation is a non-specific disorder. It also occurs in psychiatric diseases and in normal life, for instance, in cases of fatigue. Greenberg et al. (1984) reported such depersonalisation feelings being combined with the conviction of already being dead in a certain kind of epileptic seizure. These seizures, known as complex partial seizures, are often connected with damage to the temporal lobe. Lewis and Watson (1987) described a 16 year old female with such seizures originating in the right side of the posterior temporal lobe. In her case, recurrent autoscopy was the only common feature of the seizures.

As mentioned above, out-of-body experiences are more than autoscopy. Here people have a feeling of being up and seeing themselves from a bird's-eye perspective. Such experiences occur in 10 % of the whole population (Green 1968, Blackmore 1982). They are often connected with migraine, epilepsy, and schizophrenia, but also occur in psychically healthy people. There are hints that people who have the ability to experience an "exteriorized perspective" (Irwin 1986) when they dream more often report out-of-body experiences. An exteriorized perspective means that they see themselves as their own counterpart. This finding led to discussions about whether out-of-body experiences resemble dreams, whereas it is often the closeness to reality that is emphasized in out-of-body experiences.

But will this kind of comparison of the complex phenomenon "dream" with the complex phenomenon "out-of-body experience" really lead to new insights? Going against this, it seems to be important that a neuro-anatomically grounded ability of the brain could be regarded as a common basis for these different phenomena. This would also raise the question of whether people with such "abilities" would have better results than average in visual spatial tasks in neuropsychological tests.

About 60 years ago, the American scientist Wilder Penfield (born 1891 in Spokane, state of Washington, U.S.A., died 1976 in Montreal, Canada) and his colleagues carried out seminal investigations which are often quoted even today when the topic is near-death experiences. However, Penfield's primary intention was not to do research on near-death experiences, but rather to investigate brain function in people with epilepsy. In those days, the treatment of epilepsy was seriously inadequate. The wide range of antiepileptic drugs available now did not yet exist. Today most people with epilepsy can get full relief or at least sufficient relief to be able to live close to what we call a normal life. Formerly, brain surgery was often the last chance for cure, but it was not always successful. How does brain surgery work in cases of epilepsy? Firstly, brain cells with abnormal activity have to be detected by detailed encephalographic (EEG) recording. Then this area of the brain is simply removed. If such surgical intervention is too extensive, that is, if it includes large areas or certain structures, side-effects with neuropsychological symptoms are likely to occur.

An example of this are the so-called split-brain patients, in which an operation to sever the corpus callosum has been performed. The corpus callosum is a structure between the two halves of the brain which connects them together. The cut was made to attenuate extremely severe epileptic seizures which spread from one hemisphere to the other. But the operation had many side-effects, because the two halves of the brain were no longer able to communicate after the operation and this led to important neuropsychological problems.

But in the surgical interventions of interest to us here, circumscribed areas of the temporal lobe were extracted. And prior to that, Penfield carried out experiments in which he stimulated these areas with needles to provoke neurological or psychic experiences in his patients in an attempt to figure out which regions of the brain were responsible for which neurological or psychic phenomena. These experiments did much to launch the relatively young science of neuropsychology. The patients were awake. Operations were conducted using only a local anaesthetic because brain tissue itself has no receptors for pain.

One of Penfield's patients was 33 years old and suffered from temporal lobe epilepsy. Not all types of epilepsy follow the "traditional" pattern of loss of consciousness followed by convulsions. The pattern depends on the localisation of the disturbed activity in the brain. So even the pattern of seizures could tell us something about the function of that area in normal, healthy conditions. The seizures of the young man occurred in the following way. At first he heard a sudden noise or smelt something like perfume. Then he was overcome by feelings of familiarity, or fear, or a feeling of seeing himself. And that was already the whole seizure! Just before the operation, the patient reported a strange feeling that everything, even the current situation in the operating room, was something which had already taken place in the past. In short, he felt the present as past, and moreover had the feeling that he was already in the future and would now experience the past. These feelings were accompanied by sensations in the stomach and rectum. These can also be regarded as the equivalent of a seizure. The altered sense of time is interesting because it is also reported by people who have had near-death experiences in clinical death. When stimulating an area of the anterior right temporal lobe, the patient exclaimed suddenly:

Oh god! I am leaving my body. (Penfield 1955, p. 458)

Then, vestibular sensations were provoked, giving a feeling that the body was in motion. The patient thus had the feeling that he was standing up, whereas he was still lying down on the operating table. In other studies, Penfield also reported an altered sense of movement, caused by stimulation of the upper temporal lobe. This patient uttered:

I heard voices. My whole body seemed to be moving back and forth, particularly my head. (Penfield and Perot 1963, p. 651)

In addition, the lifelikeness, immediacy, familiarity, and lucidity of the experiences of those who underwent Penfield's stimulation experiments are noteworthy.

Another author (Ionășescu 1960) reported about a 32 year old patient with right temporo-occipital epileptic focus whose attacks featured a re-experiencing of scenes from the past:

> He also has the impression, while the seizure lasts, that he relives long-past schoolboy scenes and meets his former colleagues. (p. 172)

An altered sense of time, out-of-body experiences, lifelikeness, and meeting with former acquaintances—these are also phenomena which occur in near-death experiences and which are often regarded as exclusive characteristics of near-death experiences. But as we see from the above, they simply aren't! But do epileptic seizures occur in clinical death? The answer is negative! These cases merely exemplify the often mentioned theory of multiple realization of mental states, i.e., different triggers can cause the same experiences. Penfield's experiments do not explain the cause of near-death experiences in clinical death, but they do give us hints about which brain regions may be involved in such a situation.

Autoscopy and out-of-body experiences are integrative functions of the brain, as the terms themselves reveal. They have a cluster structure and they are composed of many different experiences, including at least the following:

* a visual perception: seeing a human body,
* identification of visual perception with the self, or
* remembrance of one's own mirror image,
* a feeling of movement, for instance turning around,
* a feeling of hovering.

Not all elements have to occur at the same time. This means that what we discussed regarding the term "near-death experiences" continues here, i.e., we have terms which are established from various minor terms—in our case experiences. These minor terms overlap with other mental states caused by completely different triggers. The minor terms are not required to occur all at once or even in a certain order. Furthermore, none of the above-mentioned minor terms or experiences which could belong to an out-of-body experience are absolutely essential for it! We could omit one or the other and nonetheless still use the term. All in all, out-of-body experiences in clinical death are not a dogma; they are heuristically explainable.

The structures of the brain responsible for recognition of a face, attribution of visual experiences to the self, and enabling us to rotate objects in our imagination are important in the induction of such experiences. The latter function—imagining the rotation of an object—is linked to the angular gyrus, whereas feelings of alienness or familiarity are connected with the temporal lobe structures. The angular gyrus is a region which connects the parietal lobe with the temporal and occipital lobes. It is thus assumed that its function is to integrate everything "elaborated" in the three lobes to build up new and more complex experiences.

At this point there begins another approach to the explanation of out-of-body experiences. We know about manifold abnormal experiences—quite independent of near-death experiences in clinical death—in which such integrative functions of the brain are severely altered. These include so-called delusional misidentification syndromes. People who suffer from this, for instance, take other people they see for copies of themselves. For instance, it is not their neighbour they see, but rather an identical-looking but strange person. In other subtypes of such syndromes, the situation may be like this: someone who has seen his neighbour many times would think that he has met a different person each time, despite the fact that the person had the same appearance on each occasion. Affected people therefore distrust others or certain people, because they take them for conjurers or impostors.

Another weird symptom is the alien-limb phenomenon. Affected people think that one of their limbs does not belong to their own body. They try to get rid of the strange limb, for example, by throwing their own limb out of the bed, with the consequence that they themselves fall out of bed. This symptom could have different causes. One possibility is *corticobasal degeneration*, a disease with features of Parkinson's disease, but with fatal prognosis. In the best case, it is provoked by long immobilisation of a limb in a cast, as so impressively described by Oliver Sacks (born 1933 in London) in his 1984 publication "*A leg to stand on*", in which he wrote about his own experiences with that symptom after an accident. In such cases, the prognosis is good. In this context, we should also mention the "opposite" symptom, when people have aches in a limb that no longer exists because it has been amputated. This better known phenomenon is called phantom limb. Even supernumerary phantom limbs are reported, where people *feel* additional limbs or body parts which are superfluous, i.e., they do not exist and have never in fact existed.

What conclusions can we draw from the existence of these symptoms and diseases? The normal feeling of healthy individuals that all parts of their body belong to themselves is not an "inborn" and immutable function; it is rather an integrative function of multisensory brain functions which can easily be disturbed. Certain experiments are truly fascinating. In a study by Ehrsson

(2007), healthy probands wore head-mounted displays which were connected with two video cameras behind the person's back. In these displays they saw their own back filmed by the camera behind them. Then a researcher simultaneously touched both the person's real chest and the illusory body by moving another rod in front of the camera. For the probands, the rod touching their real chest was out of sight. They imagined visually that they were witness to someone touching their back, but at the same time they felt something touching the front of their upper body. So after some repeated stimulations, the brain became disturbed. It was receiving apparently contradictory sensorial information: a feeling at the front but a view of the very same event happening at the back. The result was that the probands had the feeling of sitting behind their own body.

This is an opportune moment to say a few words about reports that people have seen things in an operating theater during an out-of-body experience which they could not possibly have witnessed. Such reports do not stand up to criticism. On the one hand, most people are well informed about the things that usually happen in an operating theater—by soap operas on the TV or movies. It is notable in this context that reports are often formulated somewhat generally. There is thus ample scope for interpretation by both experiencer and investigator, as we can see in a publication of Sabom (1981), in which personal interpretations and intentions play a prominent role. On the other hand, investigations in operating theaters under controlled circumstances, e.g., using tags in the form of slips at certain locations which could not have been detected from a lying position on the operating table, revealed no hint whatever of any extraordinary abilities (see Kasten 2008). Nevertheless, the suggestion that some people hear voices or see the lights of the operating room is quite possible and could be explained by anaesthesia awareness—a problem we shall discuss in some detail shortly.

4.3 Near-Death Experiences in Clinical Death—Result of a Severe Brain Function Disorder?

There are serious hints that the likelihood of near-death experiences does not only depend on a certain predisposition of the brain, such as the ability to see oneself as one's own counterpart in dreams or probable personality features. Moreover, the severity of the impairment of brain function in clinical death is a considerable influence factor. Martens (1994) investigated patients after cardiac arrest and found that near-death experiences occurred only in those cases where cerebral ischemia lasted rather a long time. Van Lommel et al. (2001) found that, in cardiac arrest survivors, a higher occurrence of near-death experiences is, among other things, connected with memory problems after prolonged cardiopulmonary reanimation. This finding also supports the assumption that the severity of the clinical death event increases the chances of developing near-death experiences. This is also supported by a recent study in Slovenia (Klemenc-Ketis 2010) which has given rise much discussion. They also investigated cardiac arrest survivors. One in five reported near-death experiences. Soon after reanimation, their blood was examined. Patients with near-death experiences had higher concentrations of carbon dioxide and potassium in the blood. But these findings do not imply that carbon dioxide itself would cause near-death experiences! A high amount of carbon dioxide in the blood immediately after reanimation indicates a disturbed arterio-venous relation of the blood gases oxygen and carbon dioxide, and its persistence is a sign of impaired convalescence. This implies that circulatory arrest was more severe in those patients with prolonged abnormalities in the blood. But the more severe or prolonged a circulatory arrest caused by cardiac arrest, the more severely the function of organs is impaired, and the more organs are endangered. The same applies to the brain and its functioning. In this sense, increased potassium could be a sign of cell damage.

Putting this together, cardiac arrest causes circulatory arrest and the latter causes impairment of the metabolism, that is, lack of oxygen and nutrients, especially glucose. The result is malfunction of the organs, including the brain. The longer this condition perdures, the higher the risk of irreversible damage, and eventually, death of the organs. Because the brain is the most sensitive organ with regard to these conditions, it constitutes the weakest link in the chain. When brain death occurs, a human being is declared dead. In this chain of events, clinical death is only a transitory state of short duration. The above-mentioned studies suggest that the more severely organ functions, and in particular, brain function is impaired, the more likely the occurrence of near-death experiences in clinical death survivors.

4.4 Pharmacological and Neurotransmitter Theories: Ketamine, Serotonin and Dimethyltryptamine

Not all near-death experiences are connected with a clinical death. As discussed above, this leads to conceptual difficulties, because we speak about a near-death experience without passing through a "near" or clinical death. There are also uncertainties in many reports. For example, experiences are often reported to have occurred during an operation, but without evidence that there had ever been a life-threatening situation during the operation. As already discussed, medication or drug consumption can trigger near-death experiences or they can occur in mentally healthy people.

Jansen (1990) showed that the anaesthetic ketamine can provoke near-death experiences. In a self-experiment, he administered ketamine and experienced "*...travel through a tunnel, emergence into the light, and a 'telepathic' exchange with an entity that could be described as 'God', although I have no religious beliefs and had no particular expectations on first experiencing the drug.*" (Jansen 1997a, p. 8).

Ketamine has been used since the 1970s. There are also reports of schizophrenia-like symptoms. In people who already suffer from schizophrenia, ketamine provoked exactly those symptoms which they knew from previous psychotic phases of their disease (Lahti et al. 1995). Ketamine is similar to phencyclidine (PCP), a drug which has been in use since the 1950s. Both ketamine and phencyclidine inhibit the same receptors in the brain, in particular NMDA receptors (Breier et al. 1997).

The two substances provoke similar psychic symptoms. Ketamine is even used by drug addicts. Prominent effects are visual and acoustic hallucinations, a feeling of slowing down, delusional ideas, enhancement of sexual and musical feelings, intensification of feelings (Lim 2003), out-of-body phenomena (Freese et al. 2002), and "near-death experiences" (Corazza and Schifano 2010). All the phenomena of near-death experiences even occur with standard therapeutic dosages, not only in overdose situations (Jansen 1997a; Remerand et al. 2007)! Furthermore, severe anxiety can occur, as it does in near-death experiences (Jansen 1997b), as well as motor agitation and nightmares after anaesthesia, in the awakening phase.

Formerly ketamine was used very frequently. In emergency cases, it was already administered as a pre-medication in the ambulance in order to prepare the subject for operation. However, owing to the many psychic side-effects, it is now used less frequently in anaesthesia. Instead of ketamine, its derivative esketamine is used, since it has fewer side-effects.

Studies using positron emission tomography (PET)—a kind of medical imaging which shows the function or metabolism of organs—reveal increased metabolic activity in prefrontal brain areas under administration of ketamine. This suggests that malfunction of the prefrontal region causes the flowery types of psychically abnormal experiences reported under ketamine. Many authors not only emphasize the prominent role of the prefrontal cortex, but go much further, to the level of cell receptors. Such structures are important for the transmission of signals between nerves. There are different types of receptors in the brain. Malfunction of one kind, NMDA receptors, is connected not only with memory impairment but also with psychotic phenomena such

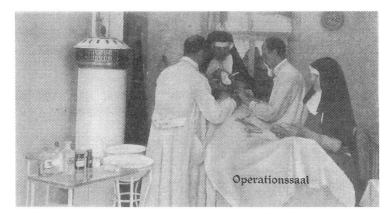

Fig. 4.3 Performance of an operation in the hospital of the Order of Saint Elisabeth in Vienna in the 1920s. The nurse in the middle is holding a mask for anaesthesia; vintage postcard

as delusion or hallucinations. There is discussion as to whether ketamine and phencyclidine could be regarded as pharmacological models of schizophrenia (Lahti et al. 2001; Morgan et al. 2009). A flaw in such pharmacological models is reductionism, because they break the multitude of symptoms down into one thing—a certain receptor or a neurotransmitter. For this reason, they do not meet the requirements of a model that could explain the complexity of brain functions and mental states. However, such models are indispensable, because they show how neurotransmitters work together with receptors. In this sense, pharmacological models are one piece in the great jigsaw of neuropsychological brain functions.

Long before ketamine, other anaesthetics were used. They also had side-effects! Dunbar (1905) described a self-experiment with chloroform (Fig. 4.3):

> It seemed to me that deep down somewhere in my consciousness voices were wrangling and quarrelling… (p. 77).

Other experiences are "*unreality of common sensations*", and "*flashing of stars in the visual field*" (p. 76).

It is interesting that substances which influence the serotonergic system, for instance, the drug dimethyltryptamine (DMT), an alcaloid of tryptamine, could provoke out-of-body experiences. Moreover, there is some discussion about whether DMT is produced naturally in the body, in the pineal gland (epiphysis cerebri) in particular, and whether it might be important for dreaming (Strassman 2000). Other speculations, such as the idea that decedents would have higher concentrations of DMT, as published some years ago in the internet encyclopedia *Wikipedia* (2007), have never been verified.

Some authors (Morse et al. 1989) postulated a certain influence of seroto-nergic structures in the temporal lobe provoking near-death experiences—a point we shall touch upon again when we discuss the model known as patho-clisis. Another author (Shawn 2004) prefers a neurotransmitter called amino-butylguanidine (agmatine) as being responsible for near-death experiences.

Furthermore, it is important to emphasize that experiences resembling those in "near-death" can be generated by drugs. Consumption of cannabis provokes out-of-body experiences, as does consumption of lysergic acid di-ethylamide (LSD). The latter also induces feelings of omnipotence. Someone who consumes LSD has the feeling of flying or being able to fly. Visions of bright light and tunnel phenomena also occur (Bates and Stanley 1985; DHS 2000).

Attempts to find a certain substance which would explain all the features of near-death experiences have to be subjected to criticism. There are serious doubts as to whether the manifold experiences constituting such an indistinct term as "near-death experiences" could be explained once and for all by the action of a single or even a small number of drugs, remedies, or neurotrans-mitters; and this against the backdrop of different triggers—not clinical death alone—which evoke similar experiences.

The various approaches discussed in this chapter reveal the complexity of interactions at the level of neurotransmitters and once again highlight the phenomenon of multiple realization of mental states. All in all, these explana-tions support critics of the term "near-death experience" who point out that the experiences subsumed under it are neither clearly assignable, nor specific to the term.

4.5 Near-Death Experiences in Other Mental States

We have already seen that experiences which are usually associated with the term "near-death experiences" can also be induced by psychotropic sub-stances. These substances include anaesthetics such as ketamine, but also il-legal drugs like cannabis and others. These additional influence factors could explain some reports of "near-death experiences" which are not linked to a "near-death" or clinical death. They may also explain the high frequency of "near-death experiences" in surveys, which is in fact extraordinarily high com-pared to the number of successful reanimations.

Trance and possession disorders could provoke out-of-body experiences, which comprise a central experience in near-death states. Moreover, illusion-ary perception of light, clairvoyance, meeting with decedents, viewing one's

Fig. 4.4 Siberian shamans, wood engraving around 1870

own skeleton by stripping of one's flesh and blood are experiences that are embedded in shaman initiation rites (Eliade 2004). Such experiences have a cultural importance. In addition, this is a point of contact with those mental states which psychiatry declares as abnormal. Whereas the "symptoms" are similar to those in other contexts, the cultural context becomes a major criterion of assessment here. Provoking such experiences is important for the work of shamans, so they get in touch with the supernatural by a shamanic flight which could be regarded as an out-of-body experience. Several procedures are used to facilitate such experiences. In the mid-19th century, the Austrian psychiatrist Max (Maximilian) Leidesdorf (born 1818, Vienna; died 1889, Vienna) mentioned the practices of Siberian shamans who entered the trance state by suffocation. They would put a rope around their neck and tighten it to ecstasiate (Leidesdorf 1865, p. 123) (Fig. 4.4).

In the central African country of Gabon, a drug called ibogaine which derives from the Iboga shrub is used for initiation ceremonies in order to cause experiences like life review, out-of-body experiences, encounters with deceased people, a feeling of gliding through a tunnel into an ulterior world, and encounters with a divine entity (Strubelt and Maas 2008). The drug has been quite well investigated because it is sometimes used for treatment of alcohol and drug addiction in Western medicine, but it can have severe cardiac side-effects too. Around the world, a wide variety of plants are used to produce drugs for shamanic ceremonies.

Furthermore, out-of-body experiences can occur in mentally healthy individuals, but also in subjects with psychiatric and neurological diseases. Even electroconvulsive therapy can induce such phenomena! According to Morse et al. (1986), traumatic brain injury or coma also induce experiences which resemble those after clinical death.

Other noteworthy influence factors are certain circumstances that arise during operations. These include so-called anaesthesia awareness. Patients do not feel any pain, but for a short time—seconds or minutes—they are in fact awake. This short time feels to them as if it were an eternity. So after the operation, they can tell astonished clinical staff about details of the operation, such as what the surgeons were talking about. If in addition out-of-body experiences are induced by medication, we can picture to ourselves how a patient might talk about this afterwards: he might well say that he had come out of his body and observed what was going on in the operating room. Theoretically, in such a situation, the out-of-body experience could also be explained by saying that the anaesthetic did not work completely, so that the patient was awake but did not feel his own body. This is an association which resembles sleep paralysis. Such anaesthesia awareness is a relatively frequent phenomenon, occurring in one out of every 1,000 cases of narcosis. Worse are conditions in which the experience of pain is not completely switched off. This happens in three out of every 10,000 narcoses. Consider a typical hospital in the German city of Leipzig with 1,700 patient beds and 43,000 patients annually. Statistically, 320 operations are conducted every week! Now we can much better imagine how often such things occur. The chances of experiencing anaesthesia awareness are rather high. Thus, iatrogenic near-death experiences are highly probable in patients who undergo an operation.

A rather recent trend is what is known as neuromonitoring. The term stands for surveillance of the depth of narcosis by means of neuro-physiological investigations. A prominent role here is played by electroencephalography (EEG). In EEG, the electrical activity of the brain is recorded—mostly in the frontal parts of the head during operations—but this gives anaesthetists enough information about how deep narcosis is and sufficient warning when patients are in danger of coming round. This should help to reduce such side-effects of operations in the future.

4.6 Near-Death Experiences in Blind People

The word does not relate to a single object, but to an entire group or class of objects. Therefore, every word is a concealed generalization. From a psychological perspective, word meaning is first and foremost a generalization. It is not difficult to see that generalization is a verbal act of thought; its reflection of reality differs radically from that of immediate sensation or perception. (Vygotsky 1987, p. 47)

The quote is from the Soviet neuropsychologist Lew Vygotsky (born 1896, Orsha, Russian Empire, today Belarus; died 1934, Moscow), who wrote his famous essay "Thought and language" in 1931. His ideas influenced neuro-psychologists of the day, and also those of subsequent generations. In short, sensations and perceptions only concern the lowest level of language acquisition. Language is shaped by thinking. This aspect is important for people who suffer from a loss of senses, whereupon they cannot receive all the sensations and perceptions of healthy people. Yet, we are able to communicate quite adequately with those having such disabilities. We have to bear in mind that language itself is not necessarily equivalent to phonetic language. Those who have lost acoustic perception can communicate by sign language, a visual spatial language, coequal to phonetic or spoken language. Even people who have lost both acoustic and visual perception—those who are both deaf and blind—can communicate via tactile stimulations.

Sensory loss is not at all as the healthy would imagine, e.g., by closing one's eyes or plugging one's ears for a while. The difference in the case of such a short time alteration is that, in blind or deaf people, the dynamic brain soon adapts to the new conditions: recollection of previous perceptions is lost. In the course of time, other senses become sharper and occupy more "space" in consciousness. So the longer blindness has lasted, the more the person loses any imagination of visual things. But *terms* referring to previously seen things or, for instance, the vision of light in the context of our own topic here of "near-death experiences", do not disappear. These terms belonging to our everyday language retain their meaning! Thus, language is not congruent with perception, but rather is only an epiphenomenon of it. This epiphenomenon is composed of many influence factors, and visual information may be one of them, but it is not necessarily so.

In this sense, "visions" of the blind in near-death experiences might be explained as follows:

* Activation of memory which still contains visual information if *blindness has developed recently.*
* Activation of non-visual memory which comprises semantically equivalent entities in people who *became blind a long time ago.*

Healthy people listening to what blind people say prefer to develop a visual imagination of the terms they use. But this is only one possible "translation" of the terms. What this implies is that we speak the same language but have different ways of imagining its referents. Generally speaking, this problem arises for anyone with some kind of constraint on one mode of perception. As we said above, the brain has an impressive ability for plasticity, so it adapts to various impairments of function in such a way that the unimpaired senses become sharper, dominate, and compensate in the best way possible for the impairment.

Investigations of people who are blind at birth or became blind in early childhood reveal an impairment of visual spatial thinking in comparison to those who have a normal ability to see (Kennedy 2006). In experiments, subjects had to touch silhouettes of well known objects or faces, sometimes in the form of a cartoon. Here blind subjects had greater difficulty than normally sighted subjects who were blindfolded. However, people who became blind late in life were much better at these tasks than normally sighted people. Those who became blind later on obviously had the benefit of a memory of former visual experiences which had not yet faded, combined with an adaptation of the brain to tactile sense developed since they had become blind.

In reports of near-death experiences, we also have to consider the distinction between people who became blind very close to birth and those who became blind late in life. As members of the "Leipzig Society of Blind and Visually Handicapped People" have told me, shortly after the onset of blindness, spatial thinking and visual memory are still in good condition. But over the course of approximately ten years, these abilities gradually diminish until they vanish completely. So once again, being blind is in no way comparable to a normally sighted person keeping their eyes closed for a while. The upshot is that terms of our language do not necessarily depend on the integrity of our five senses. When one of the senses is lacking, other modalities will prevail, and they will shape the term henceforth.

For instance, for a person with normal vision, the term "tree" comprises notions of a brown trunk with branches and green leaves. But we would not find such a predominantly visual construction of the term "tree" in blind people, especially those who have been blind since birth. After all, where would they have obtained such visual experiences? For them, this particular term might be more closely connected with acoustic or olfactorial sensations such as the rushing of leaves in the wind or the scent of blossoms, or indeed the typical scent of a pine tree, and so on. These experiences are also familiar to the normally sighted person, but they are not so prominent when they are required to imagine a tree at short notice. So the term "tree" could be established in different ways. Nevertheless, blind people and normally sighted people are able to talk about trees without communication problems. For both, the term is clearly assignable.

As already discussed, language terms comprise a cluster structure. A tree is associated with minor terms such as "big", "branches", "wood", "green", "shade", "coolness", "trunk", "bark", "leaves", "rushing of wind", and of course many more. But not all mentioned minor terms are needed in order to assign the term "tree" to an actual entity in the park. And even when two people emphasize different minor terms associated with the term "tree", they would not confuse a plague column with a tree when making an appointment in the city centre.

These thoughts are presented as background to explain near-death experiences in blind people. Relevant reports can be find in the book by Ring and Cooper (1999), but with transcendental interpretations. For instance, they tell the story of a 43 year old woman who has been blind since birth. She had her first near-death experience at the age of twelve when she suffered from appendicitis and peritonitis. Her second near-death experience occurred at the age of twenty-two following a car accident in which she suffered severe head injuries. It should be noted that the interview was conducted decades after these near-death experiences actually happened. As we have said repeatedly, this enormous temporal latency is an important influence factor for secondary interpretation of the events. If we look at the first event, we have to ask whether the woman had actually survived a clinical death—this is extremely unlikely! Time and again, we see that we have to face crucial problems with the term "near-death experiences", and our discussion recycles in a kind of merry-go-round.

But returning to the story, the woman had the impression of being outside her body. In that state, she found herself in a *"non-physical body that had a distinct form and that 'was' [...] 'like it was made of light"* (Ring and Cooper 1999, p. 24).

After a while, she saw a tall thin body lying on a metal table, but did not realise initially that it was her own. Then she felt as though she was up on the ceiling and, eventually, on the roof of the hospital building. In the following, *"she was surrounded by trees and flowers and a vast number of people. She was in a place of tremendous light, and the light [...] was something you could feel as well as see. Even the people she saw were bright."* (Ring and Cooper 1999, pp. 25–26).

She gave an interview to a support group eleven years after her near-death experience and, among other things, described her upward motion and position above the roof of the hospital (p. 45 ff.). Here is a detail from the interview as transcribed by Ring and Cooper (1999, p. 46). The interviewer asked: *"What were you aware of when you reached that point?"*

Person:	"Lights, and the streets down below, and everything. I was very confused by that."
Interviewer:	"Could you see the roof of the hospital below you?"
Person:	"Yes."
Interviewer:	"What could you see around you?"
Person:	"I saw lights."
Interviewer:	"Lights of the city?"
Person:	"Yes."
Interviewer:	"Were you able to see buildings?"
Person:	"Yeah, I saw other buildings, but that was real quick, too."

This example clarifies various points relating to the interpretation of near-death experiences. What the person claimed to have "seen" during her out-of-body experience is relatively unspecific. There is nothing special about the fact that a hospital building has a roof, and if a hospital is located in a city, other houses, people, cars, etc., should be visible from the roof. But one thing is remarkable: the world "light" occurs frequently.

It is surely uncontroversial that, on the one hand, a person who has been blind since birth has never seen houses or people before, but on the other will nevertheless possess notions of these things. These terms belong to the daily lives of blind people as they do to those of normally sighted people.

In addition to these considerations, we must also examine associations with personal views about the afterworld. The woman came from Western culture—she lived in the USA—in which such concepts are shaped by the dogmas of Christianity. Here, light plays an important role as a symbol, as it does in other religions too. Such a cultural and religious reprocessing of former near-death experiences becomes ever more likely the longer the event dates back. In this case, several years had already gone by. Besides such a symbolic interpretation of light, another explanation is worth discussing. It may be that phosphenes and photopsia occurred. In this sense, reports of "gold jewellery" also fit—all these experiences together could be regarded as a recollection of a former malfunction of the occipital lobe. In other words, malfunction of the occipital lobe structures—as might happen in clinical death—could have led to such visual experiences: bright, shiny, golden things, flashes, structured patterns in gold and light looking like spiderwebs, and so on. Could such things also happen in blind people? Could cells in the occipital lobe generate non-specific visual elements which "enable" people to experience or "see" even if they are blind? This is an interesting neuropsychological question.

In another disability—deafness—more research results exist (Critchley 1983, du Feu and McKenna 1999, Engmann 2011b). Here it can be taken for granted that prelingually deaf people sometimes have auditory experiences which resemble tinnitus cerebri or acoasms (whistling, fizzling, plonking, or droning which is experienced as being located in the head). In the blind (and in normally sighted people too!), such a theory which postulates a certain vulnerability of the occipital lobe is compatible with the notion of pathoclisis (see p. 86).

Returning yet again to the report of the blind woman, the suggestive questions of the interviewer are also to be noted. A quick examination shows that the answers were already predetermined.

4.7 Constancy or Inconstancy of Experiences

In previous sections, we have argued that near-death experiences can be matched to neuropsychological symptoms or functions. Furthermore, we have seen that there are no unique characteristics that would point out any exceptional role of those experiences in clinical death as compared to the same experiences evolving in other psychopathologically or psychically normal circumstances.

The following table gives a summary (Table 4.1).

It is striking that many experiences are somehow connected with occipital and also temporal structures. This raises the question as to whether these brain regions might be more susceptible to various triggers which make them react with neuropsychological symptoms. In clinical death, the concept of pathoclisis could explain this finding. This will be discussed in the next section. But all such attempts to associate experiences with specific regions of the brain have their limitations. Brain regions which are assumed to generate near-death experiences could only be regarded as the most significant ones, but never the *only* regions that could provoke such experiences. It is essential to take the following points into consideration:

1. Neuropsychological phenomena always have a network character. They result from an integrative performance of the brain, in the sense that different brain regions work together to "produce" a complex phenomenon in concert. From stimulation experiments or lesion studies, we know that, in patients with temporal lobe epilepsy, for instance, out-of-body experiences are often associated with temporal malfunction. But the term "out-of-body experience" also includes aspects such as autoscopy—seeing oneself from the outside as it were. Thus it seems likely not only

Table 4.1 Possible involvement of certain brain structures in the generation of near-death experiences

Experience	Neuropsychological symptoms	Relevant brain structures
Noise, ringing, humming	Acoasms, tinnitus cerebri	Temporal lobe
Long dark tunnel	Visual phenomena because of 1. central narrowing of visual field 2. central overrepresentation of fovea?	Occipital lobe, occipital pole especially
Light, sometimes with patterns (zigzag, spider-web)	Phosphenes, photopsia	Occipital lobe, intrahemispherical side especially, Brodmann area 17 and 18
Film-clip-like scenes, review of one's life, but always fragmental	Formed visual hallucinations	Occipital lobe, Brodmann area 19 or temporoparietal? (symptoms also by right temporal stimulation according to Penfield 1963)
Meeting with decedents or their ghosts, light beings	Optic/visual hallucinations	Occipital lobe? But presumably extending beyond, symptoms also by right temporal stimulation according to Penfield, 1963
Seeing oneself	Autoscopy	Temporal lobe, predominantly right (?)
To be out of one's own body, to see one's own body from a distance, suspension over one's own body	Extended autoscopy with out-of-body experience	Temporal lobe, right (?), multimodal integration of different senses—angular gyrus or temporo-parietal junction
Altered sense of time		Induced by stimulation experiments in right temporal lobe (Penfield 1955)
Feelings of love, warmth		Activation of limbic system? (or, secondary interpretation?)

that multimodal integration is altered, which leads to an incorrect feeling regarding the position of the body, but also that activation of the visual memory of the self image is altered. Everybody is able to imagine himself in his mind's eye without a mirror in front of him, in the same way that we can imagine the face of someone we know well.

2. All the considerations and reflections discussed here are only of a heuristic nature. We deduce them from well known facts, such as lesion studies or stimulation experiments. Because we know how the brain reacts to certain triggers, we assume that in other circumstances such as clinical death, the same regions must be involved. In the context of clinical death, the theory of pathoclisis belongs to such a deduction. We might also describe it as an extrapolation. We have certain well known observations which could be explained rather well, but there is one gap—in the state of clinical death we cannot measure anything. Now it is important to assess how experiences which occur in the "gap" relate to the well known phenomena. Deduction is the only way to approach a scientific explanation when we investigate a process which is impossible to measure, and clinical death is just such a process. There is no way to detect what is going on in the brain during clinical death, for instance with an fMRI scan, and this is how things should stay in the future! Clinical death is an emergency state! The priority is resuscitation—bringing a patient back to life—and nothing else. Any experiments to provide us with a few pieces of information about our philosophical interests would be unethical in such a situation. So present day research findings based on interviews shortly after awakening from reanimation, or blood tests at the same point of time, are as far as we can go toward investigating clinical death.

3. It is still unclear whether different near-death experiences after clinical death show a regular pattern or whether there are fundamental differences in all experiencers. The very question may seem surprising given that there so many reports of near-death experiences. But reports that can be unambiguously linked to clinical death are still rare, as most reports are based on interviews made years after a possible event and without proof that there really was a state of clinical death. Furthermore, as already discussed, most near-death experiences are not "*near-death*" experiences because they do not even happen in dangerous situations. Regarding the question of constancy of pattern in experiences after clinical death, two views are worth discussing:

    – Impairment of brain function differs from patient to patient, which means that different brain regions are involved in each case. This might be expected to lead to completely different near-death experiences between patients.

    – Alternatively, certain near-death experiences may dominate. Some studies suggest that visions of light and out-of-body experiences occur more often than other experiences. This implies that some brain regions may have a higher susceptibility to malfunction due to metabolic and circulatory disturbances in clinical death. This view would support the theory of pathoclisis.

4.8 The Model of Pathoclisis—A Theory of Everything?

The similarity of near-death experiences reported by people who survive clinical death could be explained by the idea that the same anatomical structures of the brain are severely affected by the circumstances of clinical death. In light of our preceding considerations, it is worth asking whether occipital and temporal malfunction might provoke most near-death experiences.

Do these parts of the brain exhibit a certain vulnerability, or pathoclisis, compared to other brain regions? Pathoclisis implies a tendency to be pathological. It refers to the inclination of an organ, a tissue, or parts within an organ to develop pathological changes under certain circumstances. These circumstances can also be manifold, e.g., alteration of the supply of nutrients or oxygen, contact with poisons, etc. The main thing is to point out that neither all organs together nor all parts of an organ together will have the same tendency to develop pathological changes. Such changes invove the following kind of temporal sequence:

disturbed functioning → loss of a few cells → loss of a larger amount of cells which leads to damage of parts of an organ, then damage of the whole organ → complete loss of organ function → death of the organ.

Pathological changes in organs express themselves through anatomical changes and malfunction. Anatomical (cellular) changes often come first, long before malfunction occurs. Every organ can compensate for disturbance to a certain degree. The most important influence factors are the time and the extent of the disturbance.

As an example that is particularly relevant to our discussion, not all brain regions have the same susceptibility to a lack of glucose or oxygen. According to Riede und Schäfer (1993), the cerebral and cerebellar cortex, parts of the striatum, thalamus, Ammon's horn, and lower olivary body of the medulla oblongata have a high vulnerability. Sacks (1996) reported a high vulnerability of the occipital lobe in relation to disturbance of colour perception.

The theory of pathoclisis was established more than eighty years ago by the French-German couple Cécile Vogt (born 1875, Annecy; died 1962, Cambridge) and Oskar Vogt (born 1870, Husum; died 1959 Freiburg/Breisgau). Oskar Vogt received his education in different hospitals where the chief physicians were notables such as Otto Binswanger (born 1852, Scherzingen; died 1929, Kreuzlingen) in Jena, Germany, Auguste Forel (born 1848, Morges; died 1931, Yvorne) in Burghölzli, Switzerland, and Paul Flechsig (born 1847, Zwickau; died 1929, Leipzig) in Leipzig. All of them were protagonists of brain research. He also worked at Salpêtrière in the University of Paris. During his stay in Jena (Germany), as an assistant in the department of anatomy, he got *"involved in his first serious anatomical studies of the human brain"* (Klatzo 2002, p. 3).

Cécile Vogt studied medicine at the University of Paris. She then worked at the Clinic at Bicêtre which was headed by the famous neurologist Pierre Marie (born 1853, Paris; died 1940, Paris). In 1899, Oskar and Cécile married. From then on they did research together at different centers and laboratories. Both were keen to investigate how psychic functions are linked with the anatomy of the brain. In 1914, Oskar Vogt became director of the Kaiser Wilhelm Brain Research Institute in Berlin-Buch, and in 1925 he also became scientific director of the Moscow Brain Research Institute, especially for investigation of Lenin's brain (Klatzo 2002). The Soviet leader and revolutionary Vladimir Ilyich Ulyanov alias Lenin (born 1870, Simbirsk/Ulyanovsk; died 1924, Gorki) was considered to be an outstanding and wise person. It was supposed that these personal features would have their counterpart in certain anatomical peculiarities of the brain. Oskar Vogt accepted the challenge to investigate this microscopically.

He described Lenin's brain as big and with a high level of pyramidal cells (Meyer 2000). He thus concluded that the large amount of pyramidal cells was the anatomical basis for Lenin's outstanding abilities—a convenient thesis for modelling an icon. But Vogt's superheated postulate that there must be some strong link between pyramidal cells and personality patterns immediately provoked counterarguments, because it smacks of reductionism. Today psychologists and psychiatrists still struggle with concepts of personality and the influence of various factors.

During their stay in Berlin-Buch, which continued after the Moscow intermezzo, in fact until 1937, the Vogt couple established their theory of pathoclisis. Experiments were performed on animals, injecting them with pharmacological agents. Furthermore, the new technique of electroencephalography was used to detect (local) electric disturbances after administration of certain agents to experimental animals. The figure and the following discussion explain how this work was performed (Fig. 4.5):

On the left, "praecgr" stands for *Area praecentralis granularis*, which is located in the upper forehead of the rabbit. On the right, "str" stands for *Area striata*, which is located on the opposite and occipital "edge" of the rabbit brain. At the top are EEG curves under normal conditions (i.e., lines 1 and 3), and at the bottom, the curves after intoxication with strychnine (i.e., lines 2 and 4). Line 2 reveals an abnormal spike and wave pattern, but line 4 remains unaffected. What did this finding reveal? The authors concluded that the *Area praecentralis granularis* (lines 1 and 2) was more susceptible to strychnine than the *Area striata* (lines 3 and 4). This revealed a certain pathoclisis of the dif-

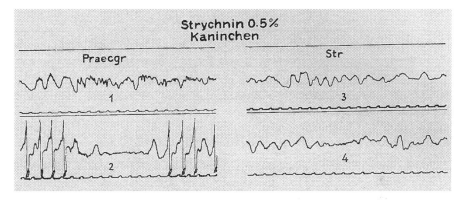

Fig. 4.5 EEG recording in a rabbit, performed by C. and O. Vogt.

ferent brain regions of the rabbit. But it is noteworthy that, with other toxic agents, the pattern of pathoclisis in the brain of a rabbit could differ. So various pathoclisis models were established, depending on the influence factor. As we can see, establishing a pathoclisis model is like piecing together a jigsaw.

Neuro-anatomical investigations were also performed on human brains. All in all, it was a painstaking task to figure out the so-called topistic areas of the brain, i.e., regions with a certain susceptibility to specific agents.

Pathoclisis theory influenced many other researchers in the growing subject of neuro-psychology. For example, the term "simple dissociation" due to Hans-Lukas Teuber (born 1916, Berlin; died 1977, Virgin Gorda), which stands for different degrees of vulnerability of complex neuro-psychic phenomena with regard to certain brain regions, seems to corroborate the Vogts' theory of pathoclisis. Teuber also coined the term "double dissociation"—see below (Teuber 1955). He worked mainly in the USA.

Today, pathoclisis is again the subject of lively debate in research about Alzheimer's disease. Degeneration of the brain occurs in reverse order compared with cerebral maturation and phylogeny. The disease starts in the temporal lobe, and in particular in the hippocampal area. From there it spreads over the limbic centres to the frontal and parietal lobes, until the whole brain is affected. Other regions of the brain such as the auditory system and to a certain degree the motor system remain unimpaired (Braak and Braak 1996). A physician named Schaffner (biographical data unknown), a contemporary of Cécile and Oskar Vogt whom they had cited (Vogt and Vogt 1922, p. 149), argued that the evolutionarily young structures of the brain had a high vulnerability. Moreover, a more recent publication by Schmitt (2005) explains such a selective vulnerability as seen in Alzheimer's with reference to certain neurotransmitters which predominate in these regions. These are serotonine, noradrenaline (i.e., norepinephrine), histamine, and acetylcholine.

We have already dealt with neurotransmitter hypotheses for near-death experiences. If certain brain regions are indeed more vulnerable in clinical death than others, the question is whether variations in the distribution of neurotransmitters could be the cause. If so, it is certainly not the only one! Lewis and Watson (1987) suggest that the posterior temporal and occipital cortices are highly susceptible to hypoxia (i.e., lack of oxygen) and hypercapnia (i.e., excess of carbon dioxide in the blood), because these cortices *"are the areas of terminal supply of the posterior (vertebrobasilar) circulation"* (p. 828). So they support the thesis that near-death experiences have their origin in the occipital and posterior temporal brain structures.

Cécile and Oskar Vogt originally hoped to provide medicine with more detailed and universal findings:

> We want to draw attention to the point that psychic disturbances which we observe in normal people at the end of a lethal disease or in agony lead back not only to circulatory disturbances in the brain, but to changes of cerebral parenchyma. It is important to analyse psychic behaviour of dying people more precisely and relate this to cerebral changes in order to elucidate the pathophysiology of premortal psychic symptoms, which is important for psychiatry too. (p. 165)

An important question is whether pathoclisis in a state of "near-death" or clinical death anticipates a so-called necroclisis in biological death which involves irreversible morphological—not only functional!—damage of brain substance (according to Rosental 1918, quoted in: Vogt and Vogt 1922). In other words, are those areas of the brain which the pathologist finds to be the most sensitive and the most easily damaged also the same areas as those whose function was altered in clinical death? Such thoughts raise other problems.

Cécile and Oskar Vogt described different models of pathoclisis. They distinguished specific pathoclisis from unspecific, and furthermore, with regard to localisation, they distinguished what they called monotopical and polytopical pathoclisis. In particular, the type of damage played an important role (Vogt and Vogt 1922). In near-death experiences associated with clinical death, the disturbance is almost always metabolic, hypoxia included, caused by a temporary arrest of blood circulation in the brain. In this context, many questions remain unanswered:

* How complex are these mechanisms in clinical death?
* Does one clinical death resemble another, or can they differ significantly from one another?
* Do these mechanisms always involve the same areas of the brain?

* Is the variability of such experiences not rather a counter-argument than a support for this assumption? On the other hand, what about the statistics suggesting that feelings of love, warmth, visions of light, and (varying from author to author) out-of-body experiences seem to dominate in near-death experiences? Do these not tend to support the theory of pathoclisis in clinical death?

And again we have to face the old problem:

* Where is the boundary between primary experience and secondary interpretation or secondary reprocessing of experiences in light of a person's cultural and religious views?

5

Opportunities and Limitations of Heuristic Methods

In many discussions about medical theories of near-death experiences, critics repeatedly point out that near-death experiences have nothing to do with epilepsy, migraine, or strokes. Moreover, those who survive clinical death and have had near-death experiences are not more or less often mentally ill or healthy than anyone else. Naturally, that is quite correct and could be upheld without exception! But what lies behind this kind of argumentation? Is the idea that near-death experiences are an unresolved issue that cannot be compared to anything else?

In the neurosciences, the current method is to conclude from lesions or other impairments of certain brain areas to the function of those areas. If we have a state like clinical death which is not "measurable", but if we would like to know which malfunctions are connected with it, than we can *deduce* which brain areas might have been affected. In this sense, the same phenomena (or experiences) in other "functional states" of the brain such as neurological or psychiatric diseases, or abnormal experiences caused by drugs, or in mentally healthy people provide us with information about which areas of the brain might be associated with experiences in clinical death. This is what we call heuristics—to conclude from the well known to the unknown. It is not as unerringly accurate and persuasive as an experiment, but the latter is impossible in clinical death.

Theoretically, a functional magnetic resonance imaging (fMRI) scan under conditions of clinical death could reveal blood circulation and positron emission tomography (PET) could reveal the metabolism of glucose in different brain regions. But why should we wish to do this? What would be the point? Firstly, as argued above, doing such MRIs and PETs during clinical death is unethical, since it would constitute a medical investigation without any benefit for the patient! Secondly, why should we wish to reinvent the wheel? Deduction is already a tried and tested method in neuroscience. In this sense, near-death experiences in clinical death are scientifically explainable. What is not explainable is the issue of the existence of the supernatural or gods. Even with more detailed investigations, that zeal to *"perceive whatever holds the world together in its inmost folds"* as Johann Wolfgang von Goethe put it in Faust, will never be fulfilled. Science and belief are different things.

If we deal with certain areas of the brain which might generate several near-death experiences, we have to consider the following. The first is the principle of "double dissociation" of brain functions (according to Teuber 1955, p. 283). Double dissociation means that, on the one hand, a region of the brain is responsible for different psychic functions, but on the other, a psychic function is also generated by the collaboration of several brain regions. The famous Soviet neuropsychologist Alexander Luria (born 1902, Kazan; died 1977, Moscow) wrote in his influential book "Higher Cortical Functions in Man":

> …individual areas of the cerebral cortex cannot be regarded as fixed centers, but rather as 'staging posts' or 'junctions' in the dynamic systems of excitation in the brain, and these systems have an extremely complex and variable structure. (Luria 1966, p. 29)

Another point is important in this context: complex neuropsychological phenomena consist of single compounds. There is not only the "out-of-body experience" or only the "light being". Earlier, we showed how all these experiences can be broken down into smaller terms. From a philosophical view, the question arises as to whether such disassembling can be continued *ad infinitum.*

> The higher mental functions may be disturbed by a lesion of one of the many different links of the functional system; nevertheless, they will be disturbed differently by lesions of different links. […] … every higher mental function, in our interpretation of this term, is composed of many links and depends for its performance on the combined working of many parts of the cerebral cortex, each of which has its own special role in the functional system as a whole. (Luria 1966, p. 71)

As it is for all neuropsychological, psychic, or psychiatric and otherwise descriptively gathered symptoms, near-death experiences are individually biased. Our scientific strategy is rather like peeling an onion. We peel off layer by layer, first leaving aside personal secondary interpretations of experiences flavoured by a person's cultural and religious views. Next, we approach the experiences themselves, in search of the basic neuropsychological symptom. But when will we have got down to the most basic phenomenon? We have to take into consideration the fact that terms like "autoscopy" or "out-of-body experience" have a cluster structure, and consist of minor terms, and that those minor terms again consist of other minor terms, and so on. Peeling this kind of onion can thus become a never-ending occupation.

Consequently, most heuristic models provide us with good explanations, but they always retain slight uncertainties when we search for an exact cause of a given phenomenon. For this reason, no fully satisfying scientific explanation of near-death experiences is likely to appear on the horizon in the near future either, and likewise, emotional discussions of near-death experiences will surely continue too.

Nevertheless, near-death experiences have no unique characteristics that would separate them from similar experiences in other mental states. We have no new facts, but the combinations and conditions in which they occur differ. Near-death experiences clearly belong to the sphere of science. At the same time, individual interpretation lies beyond the paths of science but is a legitimate request to be made against the background of one's personal philosophy of life.

6

Does a Diathesis-Stress Model of Near-Death Experiences Work?

The idea that certain personality traits or an inclination to abnormal experiences in daily life enhance the risk of developing near-death experiences under certain circumstances is supported by several publications. This is not a question of disease-mongering or stigmatising those who have had such experiences! The following explanations should convince the reader of the high variability of what is considered normal. This variability gives man individual traits, can give others new insights, and, last but not least, makes togetherness interesting. And even this seems to influence our susceptibility to develop near-death experiences.

Jones et al. (1984) looked for certain personality characteristics in people who had had near-death experiences, but they were *un*able to get a *"distinctive psychological profile"* (p. 114). In contrast, other research groups revealed that people who report having had near-death experiences had more often had dissociative (Ring and Rosing 1990; Greyson 2000) and paranormal (Greyson 2003) experiences, and this in contrast to those who had never had near-death experiences.

The term "dissociation" stands for separation of certain contents of perception and memory in everyday consciousness. The person concerned is unable to integrate certain aspects of actions and feelings and thus unable to build up a holistic self. Put another way, one's feeling of unity is disturbed. Symptoms can occur on different levels. For instance, the perception of time is modified and one's self-identity alters over the course of time. Such a person may feel that she or he was someone completely different a certain time ago. Experiences of foreignness can also occur in dissociative disorders. These include the alien-limb phenomena mentioned earlier, and also out-of-body experiences.

Today attempts to identify the cause of dissociative disorders are orientated toward the concept of conversion due to Sigmund Freud. A key postulate is that psychic traumata precede the disorder. These traumata include events such as sexual abuse and physical abuse, but also psychic conflicts which cannot be solved and finally work their way into somatic symptoms. Along these lines one has the observation of Irwin (2000) that people with near-death experiences have frequently suffered traumata in childhood. Then using the "bridge" provided by dissociation, the circle to near-death experiences would be closed.

Dissociative experiences also occur in normal daily life. The impression that everything is far away and unreal, and that one is not really present, could be described scientifically as a disorder of the boundary between the I and the surrounding world, in what is known as derealization. But this often occurs after a sleepless night and lasts only for a short time. Furthermore, self-induced dissociative states occur, as in a trance in which personal identity and exact perception of immediate surroundings are temporarily abrogated. Trance is an important part of shaman practices in many nature religions.

A shaman (see p. 98) gets into contact with supernatural beings by a trance which he induces through ritual movements, dances, or intoxicants. Out-of-body experiences also constitute such a state of trance. Even hypnosis is an induced dissociation, where the relation with the environment, but not with the hypnotist, is impaired to a high degree. Furthermore, cannabis is often a cause of dissociative states. We have to take into consideration the fact that, despite all the educational and drug prevention programs, cannabis is still one of the most often consumed drugs. In Germany, with a total population of around 82 million, 600,000 people between the ages of 18 and 64 years regularly abuse cannabis or have already been addicted to it. In addition, a report on drugs and addiction made by the Federal Government (Bundesregierung Deutschland 2009) revealed that three out of four school pupils had consumed cannabis at least once in their life. Taken together, these facts increase the statistical risk of dissociative states occurring in the population. Approximately three out of 100 people experience such a dissociative state at least once in their life.

With regard to near-death experiences, two perspectives have been discussed in the literature:

1. Affected people have personality traits which "allow" them to develop abnormal experiences such as dissociative states more often than other people who do not have such traits. If then an extraordinary burden should occur—in our discussion, a physical alteration of brain function during clinical death—easily generated dissociative states such as, for instance, an out-of-body experience may occur.
2. Near-death experiences themselves are dissociative states. That means they could be regarded as a conversion of a psychic trauma which preceded them. So they are explainable as an abnormal perceptual reaction to survival of clinical death.

The first postulate is reminiscent of the diathesis-stress model which is often used in psychology and psychiatry to explain complex disorders. The affected person already brings along a higher vulnerability as part of his constitution. The latter is shaped not only by genetic factors, but rather by early life events, or observational learning. If no other severe life events occur, the person remains healthy. But if distress occurs, a person who already has a higher burden or diathesis (vulnerability) runs a greater risk of developing psychical symptoms, disorders, or diseases. In the case of near-death experiences, this model might well be applicable to understand the interaction between personal traits and stress, but there is a danger of misinterpretation because the model is used to explain diseases or ongoing disorders, while in near-death experiences, we are usually dealing with a recollection of a previous state. Most of those who have had near-death experiences are not psychically ill.

But the kind of interaction of influence factors described in the model is supported by other findings. According to Nelson (2006), people who have hallucinations that are only connected with sleep activities more often have near-death experiences compared to those people without such hallucinations. These are called hypnagogic hallucinations if they occur when falling asleep and hypnopompic hallucinations if they occur when waking up. They are mainly visual hallucinations. Realistic, film-clip-like scenes also occur. In some cases, such hallucinations are a co-morbidity of narcolepsy, a neurological disease involving disturbed sleep patterns in which the person may suddenly fall asleep during the day or suffer from sudden muscular weakness (catalepsy). Another observation is that people who have the ability to see themselves from a bird's eye perspective in their dreams more often report out-of-body experiences than others.

Most of these studies were carried out by considering people who had had near-death experiences and asking them afterwards about certain traits, often with the help of questionnaires. In a nutshell, in these people, brain structures responsible for visual spatial imagination and rotatory imagination seem to be especially powerful. As mentioned above, rotatory imagination is linked with temporo-parietal brain structures. But there is nothing particularly special about it—everyone is different in this respect! And that shapes our individuality. There are people who cannot read a map or who have no understanding of classical painting or sculpture (in modern art, the inability to imagine something is not always due to the brain structures of the observer…). Such people may be brilliant in other modalities. They may be linguistically gifted or have musical abilities. So differences between the psychically conspicuous and psychically normal are blurred.

mentally healthy people with <u>inclination</u> to abnormal experiences

(e.g., rotatory imagination abilities in dreams, occurrence of hypnagogic and hypnopompic hallucinations, temporal EEG discharges)*

+

<u>severity</u> of clinical death

(shown by prolonged disturbance of blood gases after reanimation or a persisting malfunction of the brain, e.g., memory disorders)

=

high <u>risk</u> of developing near-death experiences in clinical death

it is not required that all influence factors must inevitably occur together

Fig. 6.1   Diathesis-stress model of near-death experiences

The above figure (Fig. 6.1) shows a concise summary of a diathesis-stress model of near-death experiences. Findings which support the model are listed in brackets. The problem is that the basic studies which delivered these findings are rare, and they often comprise rather few cases. In the sense of the model, it as also worth discussing how previous drug consumption—a high proportion of people sometimes use drugs—would also influence the upper "inclination" part of the model.

However, such a diathesis-stress model would not explain all the other experiences, such as the perception of light. Does "stress" stand for psychic stress or for the condition of clinical death when circulatory arrest leads to severe deficiency of nutrients and oxygen? A model could amalgamate different findings, but issues should not be treated dogmatically in such a model. Even when we discuss how "stress" alters brain functions or "releases" certain experiences, we come back to questions about the localisation of brain functions, thereby closing the circuit to "visions of light" and other experiences.

Concerning the second postulate which takes near-death experiences themselves for a dissociative state, many crude assumptions are involved. The exact point of time of dissociation is always in the past, while the majority of near-death experiencers who have been clinically dead are no longer suffering from a dissociative disorder at the time of the interview. Reports contain former experiences but not current complaints. Ordinarily, a patient with dissociative disorder gets symptoms repeatedly whenever triggers occur. Typical conversion theory, which derives from the work of Sigmund Freud, argues that a former conflict which has not been solved sufficiently and which is now repressed may express itself through physical symptoms. For near-death experiencers this implies that their experiences should occur frequently again and again, often released under emotional stress.

Table 6.1 Explanatory models for near-death experiences in the narrow sense, i.e., those connected with clinical death

Model/explanation	Description	Problems
Psychodynamical	Experiences result from psychic reprocessing of the traumatic event "near death". That would imply that they could be interpreted with regard to a personal background	Experiences also comprise elemental experiences such as perception of light, grid structures, and acoasms, which seem to be directly linked to a neuro-anatomical basis. An argument in favour is that near-death experiences are flowery and indeed allow interpretation of secondary reprocessed reports. Such models also involve problems with time correlation of the development of experiences
Neurotransmitters and other endogenous substances	Discussed with: N, N-Dimethyltryptamine, agmatine, opioids, and, norepinephrine (see Mobbs and Watt 2011)	Only vague assumptions, danger of reductionism to ascribe wide variety of experiences to a single substance. But neurotransmitters and other substances could indeed play a role in generation of some phenomena
Anaesthesia awareness	Narcosis insufficient; patient can perceive things happening around him to a certain degree	Not for all near-death experiences, only those which are connected with operations, but in those anaesthesia awareness seems to explain many of the experiences.
Side-effects of (anaesthetic) medicine	According to reports of side-effects of ketamine, especially, which was used as an anaesthetic until recently, and also older reports about side-effects of ether in anaesthesia	Well-established, but not all near-death experiences occur under ketamine narcosis
Pathoclisis	Certain brain regions have a higher susceptibility to lack of oxygen and glucose during circulatory arrest than other regions	Different brain regions are under discussion: temporal, occipital, or temporo-parietal junction. Problems: multiple realization of mental states, exact distinction between neuro-psychological basis phenomenon and secondary reprocessing is difficult

At this point, we have a difficult situation. The model would fit those patients who do indeed suffer from dissociative disorders and interpret their symptoms as near-death experiences, even though they have never suffered a clinical death. Alongside drug consumers, there must also be such people, otherwise we would not observe such an enormous difference between the rather small number of people who have been successfully reanimated and the considerably greater number of people who claim to have had near-death experiences. We have already expressed our concern about that issue, and this is why we wanted to speak about "real" near-death experiences—those occurring in clinical death. In this case, the second postulate does not fit, because it is not typical in these reports that people would also experience things they had "seen" in clinical death again and again, a long time after the event. Additionally, the classification of near-death experiences as a "classical" type of dissociation or a pathological perceptional reaction is not so straightforward, because then "perception" or event (i.e., clinical death) and reaction (i.e., conversion to near-death experiences) would happen completely unconsciously during coma. Furthermore, perception and reaction are rather a singular event than an ongoing psychic disorder.

The table 6.1 gives a short overview of different scientific approaches to near-death experiences.

7

Between the Poles
of Science and Religion

If we had thought at the beginning that near-death experiences would turn out to be clearly defined, well marked out, and easy to assess, comprising some specific set of experiences, we should now be struck with a profound disappointment as a result of all the above discussions. On the other hand, there is no reason to be upset by this! The very diversity of the human psyche and the overlap of neurological and psychiatric sciences with philosophy and religion make near-death experiences such a fascinating theme. The evolution of each new aspect leads to new insights and discussion points! This makes it difficult to bring together all the different aspects in a short summary. However, some points and questions crystallise out.

The sequence of experiences is neither constant within a cultural hearth nor intercultural. Some experiences seem to occur more often than others. The question is whether this finding mirrors the fact that certain brain regions are more disturbed in their function than others during clinical death and, secondly, whether these malfunctions could be the cause of such experiences. There is an inclination of certain brain regions to react more sensitively to disturbances and impairments than other regions. The lowest common denominator is that some people have been able to report bizarre experiences after a severe, reversible perturbation of brain function. None of the experiences is an essential and unique characteristic that would distinguish these experiences from similar experiences that occurred under different circumstances.

A difficult question is that of neuro-psychological basis phenomena. Do reported experiences indicate an "experienced" state as a whole, or are they a secondary interpretation? If we admit a combination of both, we have to ask to which extent each occurs. An affected person, for instance, may report out-of-body experiences because a malfunction of the temporo-parietal brain region generates experiences, or because secondary reprocessing of a life-threatening event occurs under the conviction that the soul could temporarily leave the body.

The possibility of developing experiences in clinical death seems to depend on a predisposition which includes certain personal traits and the previous occurrence of abnormal experiences in daily life, as well as the severity of brain malfunction during clinical death.

It is problematical that many distortion factors make the research and assessment of near-death experiences so difficult. Do these experiences really occur at the time of clinical death, or afterwards, during the convalescence phase? Moreover, near-death experiences also occur under circumstances that have nothing to do with a clinical death. This means that they are non-specific. Consequently, the term "near-death experience" suggests a connection of factors that does not actually exist! We should thus abandon the term "near-death experience", at least in scientific use!

To tackle the crucial problem of the manifold causes of "near-death experiences", explanations were put forward that they were a *human response to danger that can occur with or without neurological impairment* (Cant et al. 2011, p. 92) or a biologically determined pattern of experiences that could be activated on demand (i.e., in a life-threatening situation) (Schröter-Kunhardt 1993). To my mind, such conclusions are superficial because the contents of "near-death experiences" are not necessarily linked to clinical death or life-threatening situations. Furthermore, terms such as "response" or "pattern" require a certain amount of well-defined experiences which would build up the term "near-death experiences"—but we have already discussed the enormous difficulties in that field.

Another important problem is that a person can only report about their experiences a certain time after clinical death has occurred (latency). In many reports, people were interviewed years, even decades after the event! But in this context, new studies have revealed wrong or distorted recollections of daily life. The further back in time an event is situated and the more often it is recalled, the more often false memories occur. Furthermore, the intention of the interviewer is often revealed in their suggestive questions and this can alter those things the interviewed person believes they remember. The situation is the same for experiences after clinical death: Is the recollection of the content of experiences already shaped and determined by religious and cultural "foreknowledge"? Put another way, are we free to experience? An interesting question is whether religious ideas such as ascension to heaven in Christianity or Islam are an incorporation of uncommon reports that were once recounted by people who really had such extraordinary experiences. Or are they an incorporation of much older shamanic knowledge which became an inherent part of the religious concepts of the meeting of particular individuals with the supernatural? Or again, are they merely a metaphor highlighting the connection between earth and heaven?

All this taken together could be summarised under the heading of so-called secondary reprocessing, which stands for the gradual transformation

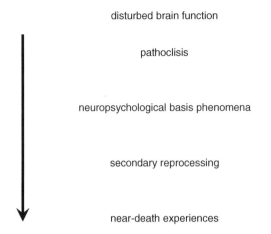

disturbed brain function

pathoclisis

neuropsychological basis phenomena

secondary reprocessing

near-death experiences

Fig. 7.1 Link between pathoclisis and near-death experiences in clinical death

of neuropsychological basis phenomena (the immediate, "real" experiences) to redesigned experiences flavoured by a person's cultural and religious views. That would imply the hypothetical process in clinical death as shown in Fig. 7.1.

Near-death experiences are accessible to rational scientific analysis and are thus scientifically explainable. The fact that most explanations are still hypothetical and tentative and that this book describes several different theories but no final solution is not a problem here. Science is always in a state of flux. On the other hand, it is important not to use science to prove the existence of gods or the supernatural as today's creationist movements currently attempt to do—movements that even want to influence the education systems in many countries! The wide variety of spiritual near-death literature can be viewed as an illustration of this. But the opposite is also true. Science cannot disprove the existence of gods or the supernatural. Science and religion are distinct entities. With regard to a survived clinical death or other life-threatening events, personal analysis of a stroke of fate could lead—as it would with other personal events—to different individual interpretations in the light of a person's worldview. Thus, personal analysis of a survived clinical death can also include spiritual or divine interpretations, just as these things could be brought to bear on all other aspects of everyday life.

The woolly term "near-death experiences" on the one hand, and increasing media coverage, on the other, could lead to a further increase in reports of near-death experiences in the future. The reason is that all kinds of experiences, including experiences after anaesthesia or drug intake, or just experiences in normal psychic life, have been subsumed under the construct "near-death experiences". And previous terms such as drug-induced psychic effects or hallucinations have received a new label: near-death experiences! The interweaving of near-death experiences with esoteric trends which have similarities to gnostic tendencies is particularly alarming. Even today, the term "near-death experiences" has already gone well beyond conditions linked with clinical death. In many studies, the term is used uncritically as a blanket term, without scrutinizing how it was established and whether all the subjects of a study can actually claim a similar basis for their experiences. These questions have to be asked before conclusions can be drawn! Alas, such tendencies seem to be an inexorable trend these days. Encounter groups are set up and self-proclaimed therapists specialise in near-death experiments. The question we should ask is whether this tendency really helps those people who have had the indescribable good fortune to survive a clinical death?

8
Important Theses

It is quite difficult to summarise in a few lines a topic which is so wide-reaching as near-death experiences. However, I shall try to highlight the most prominent points which have been—among others—a matter of discussion.

To begin with we have seen that the scientific approach to near-death experiences reveals some problems:

1. Data collection
    1.1 Most reports of near-death experiences were recorded long after they actually occurred. This results in uncertainties over whether these experiences are really linked to a clinical death, and it means that such reports will certainly have undergone secondary reprocessing in the light of a person's cultural and religious views.
    1.2 Do some experiences occur more frequently than others?
    1.3 Do basic experiences exist which reveal neuro-psychological "core" symptoms?
    The latter two questions are still a matter for discussion. Point 1.1. makes it particularly difficult to answer 1.2. and 1.3.
2. Is every clinical death always equivalent to every other clinical death? In other words, do the same biochemical changes at the level of brain function always occur in every clinical death? Or could it be that even the term "clinical death" is too broad for scientific use?
3. Science can only deliver heuristic explanatory models. Knowledge from causes of similar mental states in other circumstances is extrapolated to cases of clinical death because:
4. Near-death experiences are not directly measurable in clinical death.
5. We do not know whether near-death experiences actually occur in clinical death or rather in the phase of convalescense (see the above discussions about the value of electroencephalography in near-death research).
6. Evidence of near-death experiences caused by clinical death is unlikely to be reliable in historical reports. Those reports merely reflect metaphors to enrich a political or religious discourse. But an open question remains: Are certain contents of these reports which reveal abnormal psychic

sensations pure fantasy or are they based on real experiences which became incorporated into the reports? Those experiences did not necessarily have to derive from clinical death, but could have come from other "mental states", like psychic and neurological diseases, trance, and so on.

In contrast to these reflections, we should note the significant impact of near-death experiences on esoteric concepts. Today, near-death experiences are claimed to provide people with insights into "hidden truths". There seems to be a modern gnostic movement operating in parallel with the "traditional" religions and accompanied by a rapidly growing esoteric sector. Approximate scientific knowledge and the mingling of various scientific assumptions with concepts taken from different religions leads to pseudo-scientific explanations. The supernatural is often used in a way that could be referred to as a "god in the gap" explanation. When science comes to a point were explanation fails, this is then taken for a proof of the existence of the supernatural. Those who promote esoteric concepts usually argue against the sciences by suggesting somehow "broader" concepts. And esoteric concepts are demonstrated with the help of science by drawing upon quaint and poorly understood theories, or ones that are still the subject of controversy, often exploiting carefully selected issues. How reputable such a line of argument might be, for rejection of scientific knowledge occurs through application of scientific research findings! What a contradiction in itself! So esoteric concepts should always be scrutinized with a critical attitude, bearing in mind that the "ultimate truths" can be proven neither by scientists nor by people who claim to have deeper insights! This is the sphere of belief.

All in all, these developments reflect our time of uncertainty, in which belief and hope are gradually replaced by the need for proof and even by a form of consumerism, considering that all human endeavour, even in spiritual matters, has to be rewarded immediately. Many people find it hard to accept that man may be somehow incomplete, a mere product of evolution, and thus bound within the limitations which have shaped his development over the eons. But he who claims to have proofs of the "underlying reality" or "ultimate truths" must also subject his proofs to scientific scrutiny. Science may be incomplete too, but at least it provides society with a reliable basic language of mutual understanding.

Taking all these considerations together, my theses are as follows:

1. Near-death experiences are neither illusion nor insight. The various kinds of near-death experiences are scientifically explainable in terms of the different circumstances in which they arise.
2. The term "near-death experience" as it is used at present is unscientific and makes research difficult.
3. "Near-death experiences" are a proof neither for nor against the existence of god or the supernatural.
4. Science and religion are separate things.
5. Apart from this, as for every life event, near-death experiences also shape one's personal world view. Naturally, this may include spiritual views among other things.

9
Explanations of Scientific Terms

Acoasms Perception of noise without external trigger, acoustic hallucinations of sound and noise but without hearing voices.

Angular gyrus A brain convolution on both sides of the posterior and upper brain. Here, information from the temporal, parietal, and occipital lobes is merged and processed (see Fig. 4.1)

Aphasia A speech disturbance occurring in many different forms. For instance, words cannot be understood (= sensory aphasia), words of native language appear foreign, or words and sentences cannot be articulated any more or only with spelling mistakes and grammatical errors (expressive aphasia). Aphasia often occurs in in subjects who have had brain tumours or strokes.

Apraxia Disturbance of an organized and targeted sequence of movements; for instance, inability to use cutlery correctly for dinner or to handle a coffee machine. Apraxia corresponds to a lost function of previously learned and controlled motor abilities. It is a malfunction of the brain, often associated with strokes or brain tumours. Different forms of apraxia exist and definitions may differ from author to author.

Autoscopy (also heautoscopy) A visual, hallucinatory perception of one's own self in the visual field, similar to a mirror image. Not to be confused with the same term from otorhinopharyngology where it refers to a method of laryngoscopy!

Basal ganglia Nuclei located in the subcortex of the endbrain. They have various functions, e.g., they are important for modulating movements of the body and limbs.

Brodmann areas Zoning of the cortex or "brain map" according to histological criteria, as performed by German neuro-anatomist Korbinian Brodmann (born 1868, Liggersdorf; died 1918, Munich) at the beginning of the twentieth century. Certain functions can often be matched to certain areas, although most of this kind of association was detected long after Brodmann's work.

Cortex Outer layer of the brain, especially the endbrain (from the Latin word for bark). It consists of (a significant part of) the gray matter of the brain.

Corticobasal degeneration (CBD) or corticobasal ganglionic degeneration (CBGD) A rapid progressive neurodegenerative disease with Parkinson-like symptoms and dementia.

Dissociation Separation or failed integration of contents of perception or memory in everyday consciousness. This is the basis of hypnosis, day dreams, trance, or deep meditation. Automated processes of the body are not experienced. Dissociation is also the basis of several psychic disorders. The affected person does not act or feel like a coherent person. For instance, dissociative amnesia, derealization, depersonalisation, unconscious change of role and identity, and conversation disorder. Different models occur in depth psychology. Such dissociative states also occur under drug abuse, e.g., cannabis.

Electroconvulsive therapy Literally, electroshock therapy; a therapy in which epileptic seizures are induced artificially by current impulses. Applied in a sequence, usually over one week. Used in psychiatry to cure severe psychosis or depression which is resistant to any other therapy. Owing to the development of modern psycho-pharmacological therapy, electroconvulsive therapy is less frequent than it used to be.

Electroencephalography (EEG) Measurement and recording of changes in electrical potentials due to activity in the cerebral cortex. EEG patterns provide information about the level of excitement of the cerebral cortex, for instance, during the sleep-wake cycle. Pathological patterns are important for diagnosing certain diseases such as epilepsy or dementia syndrome, or for diagnosing brain death. Recently, there has been increasing importance in brain research and monitoring of narcoses (as part of so-called neuro-monitoring). In 1924, the German physician Hans Berger (born 1873, Neuses; died 1941, Jena) performed the first successful EEG recording on a human being, while the Liverpool physician Richard Caton (born 1842, died 1926) published papers on such electrical phenomena in animals as early as 1875.

Epiphysis cerebri (or pineal gland) An endocrine gland producing melatonin and located in the middle of the base of the brain.

Frontal Relating to the frontal lobe of the brain.

G-LOC syndrome Stands for *gravitational-force-induced loss of consciousness*. Sometimes occurs in pilots or astronauts who are exposed to high gravitational forces. Symptoms could be among others loss of colour vision ("greyout"), tunnel view, visions of lights or scenes, out-of-body experiences, and at least loss of consciousness. The cause is cerebral hypoxia.

Heuristics The art of making true statements by exploiting the similarity of unknown phenomena to well-known phenomena, using the latter to understand the former or to deduce a common origin for both phenomena. The opposite is logic – the skill of reasoning with true statements.

Hypoxia and hypercapnia Decrease (hyp-) of oxygen and increase (hyper-) of carbon dioxide in the blood. Usually measured in arterial blood as a partial pressure.

Ischaemia A lack of blood supply, involving complete cessation or deficient perfusion of an organ. In the brain, a deficient perfusion can cause a stroke.

Karma From Sanskrit, meaning "act"; a doctrine that either a psychic or physical act always has a consequence that could come into effect in one's current life or in the afterworld.

Limbic system Consists of different tracts in the brain which connect different regions. The system is involved in generating affects, emotions, and memory.

Magnetic resonance imaging (MRI) Imaging of body parts by means of a strong magnetic field. Electromagnetic waves are induced by nuclear spins and result in a detailed image of different tissues, in particular soft tissues. An fMRI (functional MRI) can show vital functions such as blood supply in real time.

NMDA receptor N-methyl-D-aspartate receptor, activated by the neurotransmitter glutamate. Ketamine and the drug phencyclidine (PCP) bind with this receptor but inhibit transmission of signals. In the brain, this receptor plays an important role in learning and memory.

Occipital Relating to the occipital lobe of the brain.

Parietal Relating to the parietal lobe of the brain.

Pathoclisis Inclination to develop pathological changes. This term describes different susceptibilities of organs or parts or zones of organs to react with tissue damage if certain harmful influence factors such as lack of blood supply or cytotoxins occur.

Phosphenes Perception of light without external trigger.

Photopsia Perception of flashes, sparks, or flickering without external trigger.

Phylogeny Historical development of individuals that share a common feature and thus are classified in a certain group.

Positron emission tomography (PET) Form of nuclear medical imaging with application of a contrast agent for imaging organs; the cutaway views provide information about the metabolism and blood supply of organs, especially concerning intensity and distribution. Used in cancer diagnostics, but also in brain research.

Posterior, vertebrobasilar blood circuit Refers to blood supply in the posterior parts of the brain, and especially in the occipital lobe. This area gets blood from both vertebral arteries, which unite to supply the basilar artery.

Serotonine A neurotransmitter occurring in the brain, but also in the intestine.

Temporal Relating to the temporal lobe of the brain.

Tinnitus cerebri Perception of noise, mostly beeping or swooshing, without external trigger, matched to the head as point of origin. This is contrasted with tinnitus aurium, where noises are matched to the ear, a very widespread disease.

References

Adam A (1969) Texte zum Manichäismus. [Texts on Manichaeism] De Gruyter, Berlin

Arnold C (1870) Die Unsterblichkeit der Seele betrachtet nach den vorzüglichsten Ansichten des klassischen Alterthums. [Immortality of the soul according to the exquisite opinions of the classical period] J.G. Wölfle'sche Universitäts-Buchhandlung, Landshut

Athappilly GK, Greyson B, Stevenson I (2006) Do prevailing societal models influence reports of near-death experiences? A comparison of accounts reported before and after 1975. J. Nerv. Ment. Dis. 194 (3): 218–222

Augustine K (2006) Hallucinatory near-death experiences. Internet Infidels 1995–2007. http://www.infidels.org/library/modern/keith_augustine/

Bates BC, Stanley A (1985) The epidemiology and differential diagnosis of near-death experience. Amer. J. Orthopsychiat. 55 (4): 542–549

Becker GW (1907) Die Geheimnisse des Todes. [Mysteries of death] Verlag Fritzsche und Schmidt, Leipzig

Beckermann A (2008) Analytische Einführung in die Philosophie des Geistes. [Analytical introduction to the philosophy of mind] De Gruyter, Berlin, New York

Belanti J, Perera M, Jagadheesan K (2008) Phenomenology of near-death experiences: a cross-cultural perspective. Transcult. Psychiatry 45 (1): 121–133

Beltz W (1987) Die Schiffe der Götter. Ägyptische Mythologie. [The ships of the gods. Egyptian mythology] Buchverlag Der Morgen, Berlin

Bhaskaran R, Kumar A, Nayar PC (1990) Autoscopy in hemianopic field. J. Neurol. Neurosurg. Psychiatry 53 (11): 1016–1017

Biedermann H (2004) Knaurs Lexikon der Symbole. [Knaur's encyclopedia of symbols] Droemersche Verlagsanstalt, München

Blackmore SJ (1982) Beyond the body. Heinemann, London

Bosing W (2012) Hieronymus Bosch. The complete paintings. Taschen, Köln

Braak H, Braak E (1996) Development of Alzheimer-related neurofibrillary changes in the neocortex inversely recapitulates cortical myelogenesis. Acta Neuropathol 92: 197–201

Bradford E (1974) Paul the traveller. Allen Lane, London

Breier A, Malhotra AK, Pinals DA, Weisenfeld NI, Pickar D (1997) Association of ketamine-induced psychosis with focal activation of the prefrontal cortex in healthy volunteers. Amer. J. Psychiatry 154 (6): 805–811

Britton WB, Bootzin RR (2004) Near-death experiences and the temporal lobe. Psychol Sci 15 (4): 254–258

Brugger P, Blanke O, Regard M, Bradford DT, Landis T (2006) Polyopic heautoscopy: Case report and review of the literature. Cortex 42 (5): 666–674

Buchner R (1956) Einleitung [introduction] In: Gregor von Tours. Zehn Bücher Geschichten, vol. 1. Rütten und Loening, Berlin

Bundesregierung Deutschland (May 2009) Drogen- und Suchtbericht [Report on drugs and addiction] Berlin, p. 64

Callaway J (1988) A proposed mechanism for the visions of dream sleep. Med. Hypotheses 26 (2): 119–124

Cant R, Cooper S, Chung C, O'Connor M (2012) The divided self: near-death experiences of resuscitated patients—a review of literature. Int Emerg Nurs. 20 (2): 88–93. doi: 10.1016/j.ienj.2011.05.005. Epub 2011 Aug 5

Carus CG (1930) Psyche. Zur Entwicklungsgeschichte der Seele. [The Psyche. Historical development of the soul] Kröner, Leipzig

Chawla LS, Akst S, Junker C, Jacobs B, Seneff MG (2009) Surges of electroencephalogram activity at the time of death: a case series. J. Palliat. Med. 12 (12): 1095–1100

Corazza O, Schifano F (2010) Near-death states reported in a sample of 50 misusers. Subst. Use Misuse 45 (6): 916–924

Critchley EM (1983) Auditory experiences of deaf schizophrenics. J. R Soc Med. 76 (7): 542–544

DHS—Deutsche Hauptstelle gegen die Suchtgefahren e.V. (2000): Drogenabhängigkeit. Eine Information für Ärzte. [Drug addiction. A resource for physicians] Hamm

Die Vorsokratiker [The Presocratics] (2010). Ed. by M. Laura Gemelli Marciano. Vol. 3. Artemis and Winkler. Mannheim

du Feu M, McKenna PJ (1999) Prelingually profoundly deaf schizophrenic patients who hear voices: a phenomenological analysis. Acta Psychiatr. Scand. 99 (6): 453–459

Dunbar E (1905) The light thrown on psychological processes by the action of drugs. In: Proceedings of the Society for Psychical Research. Trübner, London, pp. 62–77

Ehrsson HH (2007) The experimental induction of out-of-body experiences. Science. Vol. 317: 1048

Eliade M (1981) A history of religious ideas. The University of Chicago Press. Vol. 1. Chicago and London

Eliade M (1982) A history of religious ideas. The University of Chicago Press. Vol. 2. Chicago and London

Eliade M (2004) Shamanism: archaic techniques of ecstasy. Princeton University Press, Princeton

Engmann B (2011a) Bipolar affective disorder and migraine. Case history. Case Report Med: 154165. Epub 2011 Nov 28

Engmann B (2011b) Peculiarities of schizophrenic diseases in prelingually deaf persons. MMW-Fortschr Med 153 Suppl 1: 10–13

Engmann B (2011c) Cluster concept, parallelism, and heuristics in discussion of near death experiences. Journal for Philosophy and Psychiatry, March. http://www.jfpp.org/78.html

Engmann B, Turaeva M (2013) Near-death experiences in Central Asia. Advanced Studies in Medical Sciences, 1 (1): 1–10

Eusebius Pamphili (1937) Kirchengeschichte. [Ecclesiastical history] Kösel-Pustet, München

Feng Z (1992) A research on near-death experiences of survivors in big earthquake of Tangshan, 1976. Zhonghua Shen Jing Jing Shen Ke Za Zhi 25 (4): 222–225, 253–254

Fenwick P, Fenwick E (1997) The truth in the light: An investigation of over 300 near-death experiences. Berkley Trade, New York

Fichte IH (1856) Anthropologie. Die Lehre von der menschlichen Seele. [Anthropology. The doctrine of the human soul] F.A. Brockhaus, Leipzig

Flügel G (1969) Mani, seine Lehre und seine Schriften. [Mani, his doctrine and scripts] Biblio-Verlag, Osnabrück [re-print from 1862]

Fontenelle B (1989) Philosophische Neuigkeiten für Leute von Welt und für Gelehrte. [Philosophical news for the worldly and scholars] Reclam, Leipzig

Forster B, Ropohl D (1989) Rechtsmedizin [Forensic medicine] Enke, Stuttgart

Fracasso C, Aleyasin SA, Young MS (2010) Brief report: Near-death experiences among a sample of Iranian muslims. Journal of Near-Death Studies. Vol. 29 No. (1): 265–272

Freese TE, Miotto K, Reback CJ (2002) The effects and consequences of selected club drugs. Journal of Substance Abuse Treatment 23: 151–156

Freud S (1996) Die Traumdeutung. [The interpretation of dreams] Fischer, Frankfurt am Main

Freud S (2000) Studienausgabe [study edition], vol IX. Fragen der Gesellschaft, Ursprünge der Religion. Ed. by Mitscherlich A, Richards A, Strachey J. Fischer, Frankfurt am Main

Gabbard GO, Twemlow SW, Jones FC (1981) Do "near-death experiences" occur only near death? J. Nerv. Ment. Dis. 169 (6): 374–377

Green C (1968) Out-of-the-body experiences. Hamish Hamilton, London

Greenberg DB, Hochberg FH, Murray GB (1984) The theme of death in complex partial seizures. Amer. J. Psychiatry 141 (12): 1587–1589

Gregorovius F (2010) The history of the city of Rome in the middle ages. Cambridge University Press, Cambridge

Gregory of Tours (1974) The history of the Franks. Penguin, Harmondsworth

Greyson B (1983) The near-death experience scale. J Nerv Ment Dis 171: 369–375

Greyson B (1997) The near-death experience as a focus of clinical attention. J Nerv Ment Dis 185 (5): 327–334

Greyson B (2000) Dissociation in people who have near-death experiences: out of their bodies or out of their minds? Lancet 355 (9202): 460–463

Greyson B (2003) Incidence and correlates of near-death experiences in a cardiac care unit. Gen Hosp Psychiatry 25 (4): 269–276

Haedicke J (1923) Über Scheintod, Leben und Tod. Ein Beitrag zur Lehre von dem Leben und der Wiederbelebung. [On the state of apparent death, life, and death] Verlag Kultur und Gesundheit GmbH, Ober- Schreiberhau

Heim A (1892) Notizen über den Tod durch Absturz. [Notes on fatal falls] In: Jahrbuch des Schweizer Alpenclub. Vol. 27. Bern: 327–337

Heinzelmann M (2001) Gregory of Tours. History and society in the sixth century. Cambridge University Press, Cambridge

Henne-Am Rhyn O (1881) Das Jenseits. Kulturgeschichtliche Darstellung der Ansichten über Schöpfung und Weltuntergang, die andere Welt und das Geisterreich. [The afterworld. A cultural-historical description of views of creation and doomsday, the other world, and the realm of ghosts]. Otto Wigand. Leipzig

Hoepner R, Labudda K, May TW, Schoendienst M, Woermann FG, Bien CG, Brandt C (2013) Ictal autoscopic phenomena and near-death experiences: a study of five patients with ictal autoscopies. J Neurol. 260 (3): 742–149. doi: 10.1007/s00415-012-6689-x.

Homer (1962) Odyssey. Anchor, New York

Hufeland CW (1791) Über die Ungewissheit des Todes und das einzige untrügliche Mittel, sich von seiner Wirklichkeit zu überzeugen. [About uncertainty of death and the only way to prove it] Weimar

Hufeland CW (1808) Der Scheintod. [State of apparent death] Berlin

IANDS (2011) http://iands.org/research/important-research-articles/698-greyson-nde-scale.html

Ionăşescu V (1960) Paroxysmal disorders of the body image in temporal lobe epilepsy. Acta psychiatrica et neurological Scandinavica. 35: 171–181

Irwin H (1986) Perceptual perspectives of visual imagery in OBEs, dreams and reminiscence. Journal of the Society for Psychical Research 53: 210–217

Irwin H (1987) Images of Heaven. Parapsychology Review 18 (1): 1–4

Irwin H (2000) The disembodied self: An empirical study of dissociation and the out-of-body experience. Journal of Parapsychology 64 (3): 261–277

Irwin H, Bramwell B (1988) The devil in heaven: A near-death experience with both positive and negative facets. Journal of Near-Death Studies 7 (1): 3843

Jansen KLR (1990) Neuroscience and the near-death experience: roles for the NMSA-PCP receptor, the sigma receptor and the endopsychosins. Med Hypotheses 31 (1): 25–29

Jansen KLR (1997a) The ketamine model of the near-death experience: a central role for the N-methyl-D-aspartate receptor. J Near Death Stud 16 (1): 5–26

Jansen KLR (1997b) Response to commentaries on "The ketamine model of the near-death experience: a central role for the N-methyl-D-aspartate receptor." J Near Death Stud 16 (1): 79–95,

Jones FC, Gabbard GO, Twemlow SW (1984) Psychological and demographic characteristics of persons reporting out-of-body experiences. Hillside J Clin Psychiatry 6 (1): 105–115

Jung CG (1981) The Archetypes and the Collective Unconscious. In: Collected Works Vol. 9 Part 1. Bollingen, Princeton, N.J.

Kant I (1953) The critique of judgement. Clarendon Press. Oxford (1953), reprinted 2010, Oxford University Press

Kasten E (2008) Die irreale Welt in unserem Kopf. Halluzinationen, Visionen, Träume. [The unreal world in our head. Hallucinations, visions, dreams] Ernst-Reinhardt-Verlag, München

Kasten E, Geier J (2009) Neurobiological theories on near-death experiences. Z Med Psychol 18: 13–27

Kellehear A (1993) Culture, biology, and the near-death experience. A reappraisal. J. Nerv. Ment. Dis. 181 (3): 148–156

Kennedy JM (2006) How the blind draw. Scientific American, (16): 44–51 doi:10.1038/scientificamerican0906-44sp

Kim SM, Park CH, Intenzo CM, Zhang J (1993) Brain SPECT in a patient with post-stroke hallucination. Clin. Nucl. Med. 18 (5): 413–416

Klatzo I (2002) Cécile and Oskar Vogt: The visionaries of modern neuroscience. Springer, Wien, New York

Klemenc-Ketis Z, Kersnik J, Grmec S (2010) The effect of carbon dioxide on near-death experiences in out-of-hospital cardiac arrest survivors: a prospective observational study. Crit. Care 14 (2): R56. Epub Apr. 8, 2010

Kolb B, Whishaw IQ (1990) Fundamentals of human neuropsychology, W.H. Freeman and Company, New York, Oxford

Kölmel HW (1993) Visual illusions and hallucinations. Baillieres Clin. Neurol. 2 (2): 243–264

Kreps JI (2009) The search for muslim near-death experiences. Journal of Near-Death Studies. Vol. 28 No. 2, Winter 2009, 67–86

Lahti AC, Holcomb HH, Medoff DR, Tamminga CA (1995) Ketamine activates psychosis and alters limbic blood flow in schizophrenia. Neuroreport 6 (6): 869–872

Lahti AC, Weiler MA, Michaelidis T, Parwani A, Tamminga CA (2001) Effects of ketamine in normal and schizophrenic volunteers. Neuropsychopharmacology. 25 (4): 455–467

Le Goff J (1996) Saint Louis. Èditions Gallimard, Paris

Leidesdorf M (1865) Lehrbuch der psychischen Krankheiten [Textbook of psychic diseases]. Enke, Erlangen

Lessing GE (1991) Nathan the wise, Minna von Barnhelm, and other plays and writings. Continuum, New York

Lewis DW, Watson ME (1987) Explaining the phenomena of near-death experiences. Amer. J. Dis. Child 141 (8): 828

Lim DK (2003) Ketamine associated psychedelic effects and dependence. Singapore Med. J. 44 (1): 31–34

Lindley J, Bryan S, Conley B (1981) Near-death experiences in a Pacific north-west American population: the Evergreen study. Anabiosis—The Journal for Near-Death Studies (1): 104–124

Locke J (1805) An essay of concerning human understanding, vol. 1. Bye and Law, London

Loftus EF (1997) Creating false memories. Scientific American, 277: 70–75

Luria AR (1966) Higher cortical functions in man. Tavistock Publications. London

Martens PR (1994) Near-death experiences in out-of-hospital cardiac arrest survivors. Meaningful phenomena or just fantasy of death? Resuscitation 27 (2): 171–175

Meyer B (2000) Oskar Vogt seziert das Gehirn Lenins. Ein Kapitel deutscher Außenpolitik in der Weimarer Republik. [Oskar Vogt performs an autopsy of Lenin's brain. A chapter of German foreign policy of Weimar Republic] In: Edition Luisenstadt, Berlinische Monatsschrift, 6: 137–141, http://www.luise-berlin.de/bms/bmstxt00/0006gesb.htm#seite139

Mobbs D, Watt C (2011) There is nothing paranormal about near-death experiences: how neuroscience can explain seeing bright lights, meeting the dead, or being convinced you are one of them. Trends Cogn Sci 15 (19): 447–449

Moody RA (1975) Life after life. Mockingbird books, Covington, Georgia

Morgan CJ, Huddy V, Lipton M, Curran HV, Joyce EM (2009) Is persistent ketamine use a valid model of the cognitive and oculomotor deficits in schizophrenia? Biol. Psychiatry 65 (12): 1099–1102

Morse M, Perry P (1992) Transformed by the Light: The powerful effect of near- death experiences on people's lives. Villard Books, New York

Morse M, Castillo P, Venecia D, Milstein J, Tyler DC (1986) Childhood near-death experiences. Amer. J. Dis. Child 140 (11): 1110–1114

Morse M, Venecia D, Milstein J (1989) Near-death experiences: A neurophysiologic explanatory model. Journal of Near-Death Studies 8 (1): 45–53

Mylius K (ed) (1978) Älteste indische Dichtung und Prosa. [Oldest Indian poetry and prose] Verlag Philipp Reclam jun. Leipzig

Nahm M, Nicolay J (2010) Essential features of eight published Muslim near-death experiences: an addendum to Joel Ibrahim Kreps's "The search for Muslim near-death experiences". Journal of Near-Death Studies, 29 (1), Fall 2010. 255–263

Necker-Curchod S (1790) Des inhumations précipitées. Impr. royale, Paris

Nelson KR, Mattingly M, Lee SA, Schmitt FA (2006) Does the arousal system contribute to near-death experience? Neurology 66 (7): 1003–1009

Pasricha S (1993) A systematic survey of near-death experiences in South India. Journal of Scientific Exploration 7: 161–171

Pasricha S (1995) Near-death experiences in South India: A systematic survey. Journal of Scientific Exploration 9: 79–88

Pasricha S, Stevenson I (1986) Near-death experiences in South India. Journal of Nervous and Mental Disease 174: 165–170

Penfield W (1955) The twenty-ninth Maudsley lecture: the role of the temporal cortex in certain psychical phenomena. Journal of Mental Science 101: 551–456

Penfield W, Perot P (1963) The brain's record of visual and auditory experience. Brain 86: 595–696

Petersen JM (1984) The dialogues of Gregory the Great in their late antique cultural background. Pontifical Institute of Mediaeval Studies, Toronto

Plato (2006) The Republic. Yale University Press, New Haven, London

Plinius Secundus, Gaius (1994). Naturkunde [natural history]. Artemis und Winkler. Zurich

Pontrandolfo A, Rouveret A, Cipriani M (2011) The painted tombs of Paestum. Pandemos. Paestum

Remerand F, Couvret C, Pourrat X, Le Tendre C, Baud A, Fusciardi J (2007) Prevention of psychedelic side effects associated with low dose continuous intravenous ketamine infusion. Therapie 62 (6): 499–505

Riede U-N, Schäfer H-E (1993) Allgemeine und spezielle Pathologie [general and special pathology] 3rd ed. Thieme, Stuttgart, New York

Ring K (1980) Commentary on "The reality of death experiences: a personal perspective" by Ernst A. Rodin. J. Nerv. Ment. Dis. 168 (5): 273–274

Ring K, Cooper S (1999) Mindsight. Near-death and out-of-body experiences in the blind. William James Center for Consciousness Studies at the Institute of Transpersonal Psychology, Palo Alto

Ring K, Rosing CJ (1990) The Omega project: An empirical study of the NDE-prone personality. Journal of Near-Death Studies 8 (4): 211–239

Russell B (1969) History of Western philosophy. George Allen and Unwin, London

Sabom MB (1981) Recollections of death. Harper and Row, New York

Sacks O (1984) A leg to stand on. Touchstone books, New York

Sacks O (1996) The case of the colorblind painter. In: An anthropologist on Mars. Vintage, New York

Schmied-Knittel I (2006) Nahtod-Erfahrungen. [Near-death experiences] In: Traumland Intensivstation. [Dreamland intensive care unit]. Ed. by Thomas Kammerer. Books on demand. Norderstedt. 231–252

Schmitt HP (2005) Neuro-modulation, aminergic neuro-disinhibition and neuro-degeneration. Draft of a comprehensive theory for Alzheimer disease. Med. Hypotheses 65 (6): 1106–1119

Schröter-Kunhardt M (1993) Das Jenseits in uns. [The afterworld in us] Psychologie heute, 6: 64–69

Schürmeyer W (1923) Hieronymus Bosch. Piper, München

Schwabe C (1834) Das Leichenhaus in Weimar. [The morgue of Weimar] Voss, Leipzig

Schwaninger J, Eisenberg PR, Schechtman KB, Weiss AN (2002) A prospective analysis of near-death experiences in cardiac arrest patients. Journal of Near-Death Studies 20 (4): 215–232

Shawn T (2004) Agmatine and near-death experiences. http://www.neurotransmitter. net/neardeath.html

Snessarew G (1976) Unter dem Himmel von Choresm. [Under the sky of Khwarezm]. FA Brockhaus, Leipzig

Splittgerber F (1866) Schlaf und Tod. [Sleep and death] Julius Fricke, Halle/Saale

Störig HJ (1958) Kleine Weltgeschichte der Philosophie. [Brief history of philosophy of the world] Europäischer Buchklub. Stuttgart, Zürich, Salzburg

Strassman R (2000) DMT—The spirit molecule: A doctor's revolutionary research into the biology of near-death and mystical experiences. Park Street Press, Rochester/Vermont

Strubelt S, Maas U (2008) The near-death experience: a cerebellar method to protect body and soul. Lessons from the Iboga healing ceremony in Gabon. Altern Ther Health Med, 14 (1): 30–34

Teuber HL (1955) Physiological psychology. Annu Rev Psychol 6: 267–296

The dialogues of Saint Gregory the Great (2010) Ed. by Edmund G. Gardner. Evolution Publishing, Mechantville/New Jersey

The Meaning of the Holy Qur'ān. (2004) Amana Publications, Beltsville, Maryland

The Oxford Illustrated Bible (King James version) (1993) Oxford University Press, Oxford

Thonnard M, Schnakers C, Boly M, Bruno MA, Boveroux P, Laureys S, Vanhaudenhuyse A (2008) Near-death experiences: fact and fancy. Rev. Med. Liege 63 (5–6): 438–444

Thonnard M, Charland-Verville V, Brédart S, Dehon H, Ledoux D, Laureys S, Vanhaudenhuyse A (2013) Characteristics of near-death experiences memories as compared to real and imagined events memories. PLoS One. 2013;8(3):e57620. doi: 10.1371/journal.pone.0057620. Epub 2013 Mar 27.

Trepel M (1999) Neuroanatomie. [neuro-anatomy] Urban und Fischer. München, Stuttgart, Jena, Lübeck, Ulm

Twain M (2010) The innocents abroad. Wordsworth, Ware/Hertfordshire

van Hasselt, AWM (1862) Die Lehre vom Tode und Scheintode. [The science of death and apparent death] Verlag Friedrich Vieweg und Sohn, Braunschweig

van Lommel P, van Wees R, Meyers V, Elfferich I (2001) Near-death experience in survivors of cardiac arrest: a prospective study in the Netherlands. Lancet 358 (9298): 2039–2045

Vogt C, Vogt O (1922) Erkrankungen der Großhirnrinde im Lichte der Topistik, Pathoklise und Pathoarchitektonik. [Diseases of the cerebral cortex with regards to topistic, pathoclisis, and pathoarchitectonic] In: Forel A, Vogt C, Vogt O (ed) Journal für Psychologie und Neurologie. Vol. 28. Johann Ambrosius Barth, Leipzig

Vogt C, Vogt O (1937) Sitz und Wesen der Krankheiten im Lichte der topistischen Hirnforschung und des Variierens der Tiere. [Location and character of diseases in the light of topistic brain research and variation of animals] Vol. 1. Johann Ambrosius Barth, Leipzig

von Goethe JW (1962) Faust. Anchor, New York

von Kirchmann JH (1870) [commentary] In: Plato's Staat. Uebersetzt von Friedrich Schleiermacher und erläutert von J. H. v. Kirchmann. Verlag von L. Heimann, Berlin

Vygotsky LS (1987) Problems of general psychology: Thinking and speech. Plenum Press, New York

Wærferths of Worcester (1965) Dialoge Gregors des Großen. [Dialogues of Gregory the Great] ed. by Hans Hecht. Wissenschaftliche Buchgesellschaft, Darmstadt

Weber S (1916) Einführungen und Anmerkungen [Introductions and remarks to the new testament] In: Das Neue Testament unseres Herrn Jesus Christus. Herdersche Verlagshandlung. Freiburg im Breisgau

Welburn A (1998) Mani, the angel and the column of glory. An anthology of Manichaean texts. Floris books, Edinburgh

Wikipedia: Dimethyltryptamin, as of 19 July 2007. http://de.wikipedia.org/wiki/Dimethyltryptamin

Zamboni G, Budriesi C, Nichelli P (2005) Seeing oneself: a case of autoscopy. Neurocase 11 (3): 212–215

Zimmer H (1980) Die indische Weltmutter. Aufsätze [The Indian mother of the world. Essays] Ed. by Friedrich Wilhelm. Insel-Verlag, Frankfurt/Main

Figure Sources

Fig. 2.2, p. 5, Schürmeyer W (1923) Hieronymus Bosch. Piper, München. Front page

Fig. 2.17, p. 40, Schwabe C (1834) Das Leichenhaus in Weimar. [The morgue of Weimar] Voss, Leipzig; plate II

Fig. 2.15, p. 38, Antoine-Wiertz Museum, Brussels, taken from catalogue published around 1900

Fig. 2.19, p 43, Möbius PJ (1905) Im Grenzlande. [In the borderland] Verlag von Johann Ambrosius Barth, Leipzig; titel page

Fig. 3.5, p. 64, www.springerimages.com/Images/RSS/1-10.1007_s00112-010-2183-7-2

Fig. 4.1, p. 84, www.springerimages.com/Images/MedicineAndPublicHealth/1-10.1007_174_2011_280-0

Fig. 4.2, p. 85, www.springerimages.com/Images/MedicineAndPublicHealth/1-10.1007_174_2011_280-0

Fig. 4.5, p. 109, Vogt C, Vogt O (1937) Sitz und Wesen der Krankheiten im Lichte der topistischen Hirnforschung und des Variierens der Tiere. [Location and character of diseases in the light of topistic brain research and variation of animals] Verlag von Johann Ambrosius Barth, Leipzig; p. 64

All the other illustrations are either engravings or historical postcards from private archive of the author or photos made by the author.

Index